Proof Analysis

This book continues from where the authors' previous book, *Structural Proof Theory*, ended. It presents an extension of the methods of analysis of proofs in pure logic to elementary axiomatic systems and to what is known as philosophical logic. A self-contained brief introduction to the proof theory of pure logic that serves both the mathematically and philosophically oriented reader is included. The method is built up gradually, with examples drawn from theories of order, lattice theory and elementary geometry. The aim is, in each of the examples, to help the reader grasp the combinatorial behaviour of an axiom system, which typically leads to decidability results. The last part presents, as an application and extension of all that precedes it, a proof-theoretical approach to the Kripke semantics of modal and related logics, with a great number of new results, providing essential reading for mathematical and philosophical logicians.

SARA NEGRI is Docent of Logic at the University of Helsinki. She is the co-author of *Structural Proof Theory* (Cambridge, 2001, with Jan von Plato) and she has published a number of papers on mathematical and philosophical logic.

JAN VON PLATO is Professor of Philosophy at the University of Helsinki. He is the author of *Creating Modern Probability* (Cambridge, 1994) and the co-author (with Sara Negri) of *Structural Proof Theory* (Cambridge, 2001), and he has published a number of papers on logic and epistemology.

Proof Analysis

A Contribution to Hilbert's Last Problem

SARA NEGRI
JAN VON PLATO
University of Helsinki

CAMBRIDGE
UNIVERSITY PRESS

CAMBRIDGE
UNIVERSITY PRESS

University Printing House, Cambridge CB2 8BS, United Kingdom

Published in the United States of America by Cambridge University Press, New York

Cambridge University Press is part of the University of Cambridge.

It furthers the University's mission by disseminating knowledge in the pursuit of education, learning and research at the highest international levels of excellence.

www.cambridge.org
Information on this title: www.cambridge.org/9781107417236

First published 2011
First paperback edition 2014

A catalogue record for this publication is available from the British Library

Library of Congress Cataloguing in Publication data
Negri, Sara, 1967–
Proof analysis : a contribution to Hilbert's last problem / Sara Negri, Jan von Plato.
p. cm.
Includes bibliographical references and index.
ISBN 978-1-107-00895-3 (hardback)
1. Proof theory.   I. Von Plato, Jan.   II. Title.
QA9.54.N438   2011
511.3´6 – dc23   2011023026

ISBN 978-1-107-00895-3 Hardback
ISBN 978-1-107-41723-6 Paperback

Contents

Preface

Proof theory, one of the two main directions of logic, has been mostly concentrated on pure logic. There have been systematic reasons to think that such a limitation of proof theory to pure logic is inevitable, but about twelve years ago, we found what appears to be a very natural way of extending the proof theory of pure logic to cover also axiomatic theories. How this happens, and how extensive of our method is, is explained in this book. We have written it so that, in principle, no preliminary knowledge of proof theory or even of logic is necessary.

The book can be profitably read by students and researchers in philosophy, mathematics, and computer science. The emphasis is on the presentation of a method, divided into four parts of increasing difficulty and illustrated by many examples. No intricate constructions or specialized techniques appear in these; all methods of proof analysis for axiomatic theories are developed by analogy to methods familiar from pure logic, such as normal forms, subformula properties, and rules of proof that support root-first proof search. The book can be used as a basis for a second course in logic, with emphasis on proof systems and their applications, and with the basics of natural deduction and sequent calculus for pure logic covered in Part I, Chapter 2, and Part II, Chapter 6.

A philosopher who seeks the general significance of the work should be able to see in what sense it contributes to the solution of a fascinating recently discovered *last problem of Hilbert* that belongs to proof theory. The much later *Hilbert programme* had more specific aims. It is remarkable how many of the original aims of this foundational programme can be carried through in, say, algebra and geometry, and indeed in many parts of mathematics that do not involve the natural numbers and the incompleteness of their theory.

Mathematically oriented readers should be able, after a study of this book, to produce independent work on the application of the method of proof analysis in their favourite axiomatic theories.

The fourth part, on non-classical logics, is mainly aimed at the student and specialist in philosophical logic. It presents in a systematic form, building on the previous parts, a proof theory of non-classical logics, with semantical aspects incorporated through what are known as labelled logical calculi. The fundamental idea here is very clear: the various systems of modal and other non-classical logics can usually be characterized by some key properties, expressed as conditions in the standard Kripke semantics. These conditions are, taken abstractly, axioms for the frames of the said semantics, and they convert into rules that extend an underlying sequent calculus. The execution of this idea in the fourth part builds, by way of the method, on virtually everything that has been presented in the previous parts. It was a great surprise to the authors when the first of them discovered the application of proof analysis to non-classical logics in 2003, and many results in Part IV are new. This part is also useful for the study of logic in computer science. Recent years have seen a growth of literature in computer science on logical systems of knowledge presentation that stems from epistemic logic as developed by philosophers, and to which systems the method of Part IV can be fruitfully applied.

Hilbert's enigmatic last problem that decorates our title is explained in our Prologue that begins the book. The structure of the book is explained later in Section 1.3, after which a summary of the individual chapters follows. Finally, a word about what is not included: we have decided to, by and large, present our approach and let it speak for itself. Of the different parts of proof theory, we have a lot to say about structural proof theory, the topic of our previous book published in 2001. Other topics, such as the proof theory of arithmetic, ordinal proof theory, and what Anne Troelstra calls interpretational proof theory in his *Basic Proof Theory*, remain largely untouched. Troelstra's book can be consulted for a first look at these different aspects to proof theory. There is no easy introduction to the proof theory of arithmetic, but Takeuti's *Proof Theory*, especially in its early chapters, is fairly accessible. The recent book by Pohlers (2009) on ordinal proof theory is a hard read. Kohlenbach's (2008) hefty tome collects together an enormous amount of results that belong to interpretational proof theory.

This book began with a series of lectures titled 'Five Lectures on Proof-Analysis' that the first author gave in Dresden in 2003. A second series was given in Munich, and a third in Braga, Portugal, in 2006. The year after, the second author gave a more extensive course on the topic at the University of Helsinki. We thank those involved, organizers, colleagues, and students, for these opportunities. In particular, we thank Roy Dyckhoff as well as our students Bianca Boretti, Annika Kanckos, and Andrea Meinander, who have all done research that has affected our presentation. We have also benefited from comments by Michael von Boguslawski and Sergei Soloviev. All the while, we have been surrounded by the patient wondering of Daniel, Stella, and Niclas.

In 1998, when the program of proof analysis was launched by our joint article 'Cut Elimination in the Presence of Axioms' in *The Bulletin of Symbolic Logic*, we received from Jussi Ketonen the following reaction: 'I suspect that this type of work will eventually lead to a completely new kind of understanding of proofs – not only as applications of rules, axioms, or ideologies, but as a branch of mathematics'. Now, twelve years later, we hope to have realized at least a beginning of that vision.

Prologue: Hilbert's last problem

David Hilbert presented his famous list of open mathematical problems at the international mathematical congress in Paris in 1900. First in the list was Cantor's continuum problem, the question of the cardinality of the set of reals numbers. The second problem concerned the consistency of the arithmetic of real numbers, i.e., of analysis, and so on. These problems are generally recognized and have been at the centre of foundational research for a hundred years, but few would be able to state how Hilbert's list ended: namely with a 23rd problem about the calculus of variations – or so it was thought until some years ago, when German historian of science Rüdiger Thiele found from old archives in Göttingen some notes in Hilbert's hand that begin with:

As a 24th problem of my Paris talk I wanted to pose the problem: criteria for the simplicity of proofs, or, to show that certain proofs are simpler than any others. In general, to develop a theory of proof methods in mathematics.

The 24th problem thus has two parts: a first part about the notion of simplicity of proofs, and a second one that calls for a theory of proofs in mathematics. Just as the problems that begin the list, what we call *Hilbert's last problem* has been at the centre of foundational studies for a long time.

When Hilbert later started to develop his *Beweistheorie* (proof theory), its aims were much more specific than the wording of the last problem suggests: he put up a programme the aim of which was to save mathematics from the threat of inconsistency, by which one would also 'solve the foundational problems for good'.

Gerhard Gentzen, a student of Paul Bernays with whom Hilbert was working, set as his objective in the early 1930s 'to study the structure of mathematical proofs as they appear in practice'. He presented the general logical structure of mathematical proofs as a system of rules of proof by which a path is built from the assumptions of a theorem to its conclusion. Earlier formalizations of logic had given a set of axioms and just two rules of inference. Another essential methodological novelty in Gentzen's work was that he presented proofs in the form of a tree instead of a linear succession from the given assumptions to the claim of a proof. Each step in a proof determined a subtree from the assumptions that had been made down to that point, and these parts could be studied in isolation. Most importantly, such parts of the overall proof could be combined in new ways, contrary to the earlier linear style of proof. Gentzen was able to give for proofs in pure logic – that is, without any mathematical axioms – combinatorial transformations that brought these proofs into a certain direct form. Questions such as the consistency and decidability of a system of rules of proof could then be answered.

It has been generally thought that Gentzen's analysis of the structure of proofs cannot be carried through to perfection outside pure logic. This book aims at presenting a method in which mathematical axioms are converted into systems of rules of proof and the structure of mathematical proofs analyzed in the same way as Gentzen analysed proofs in pure logic. The overall aim is to gain a mastery over the combinatorial possibilities offered by a system of mathematical axioms. As a rule, such a complete mastery of the workings of an axiom system cannot perhaps be achieved. Our aim is to try to make a positive contribution to Hilbert's last problem by a gradual development of 'proof methods in mathematics', inspired by the methods of structural proof theory and illustrated by examples drawn mainly from the elementary axiomatics of algebra and geometry, and from what are known as systems of non-classical logic.

1 | Introduction

We shall discuss the notion of proof and then present an introductory example of the analysis of the structure of proofs. The contents of the book are outlined in the third and last section of this chapter.

1.1 The idea of a proof

A proof in logic and mathematics is, traditionally, a deductive argument from some given assumptions to a conclusion. Proofs are meant to present conclusive evidence in the sense that the truth of the conclusion should follow necessarily from the truth of the assumptions. Proofs must be, in principle, communicable in every detail, so that their correctness can be checked. Detailed proofs are a means of presentation that need not follow in any way the steps in finding things out. Still, it would be useful if there was a natural way from the latter steps to a proof, and equally useful if proofs also suggested the way the truths behind them were discovered.

The presentation of proofs as deductive arguments began in ancient Greek axiomatic geometry. It took Gottlob Frege in 1879 to realize that mere axioms and definitions are not enough, but that also the logical steps that combine axioms into a proof have to be made, and indeed can be made, explicit. To this purpose, Frege formulated logic itself as an axiomatic discipline, completed with just two rules of inference for combining logical axioms.

Axiomatic logic of the Fregean sort was studied and developed by Bertrand Russell, and later by David Hilbert and Paul Bernays and their students, in the first three decades of the twentieth century. Gradually logic came to be seen as a formal calculus instead of a system of reasoning: the language of logic was formalized and its rules of inference taken as part of an inductive definition of the class of formally provable formulas in the calculus.

Young Gerhard Gentzen, a student of Bernays, set as his task in 1932 to develop a system of logic that is as close as possible to the actual proving of theorems in mathematics. His basic observation was that reasoning in mathematics uses assumptions from which conclusions are drawn. Some steps of reasoning analyse the assumptions into their components, others move from these components towards a sought-for conclusion. The two-way rules of such reasoning make up a system known as **natural deduction** that has only rules of inference, but no logical axioms at all. This change from axiomatic to rule-based systems marks a break with the existing axiomatic tradition as upheld by Hilbert and Bernays. Each form of logical expression, say a conjunction $A\&B$ ('*A* and *B*') or an implication $A \supset B$ ('if *A*, then *B*'), has a rule that gives the sufficient conditions for inferring it: to infer $A\&B$, it is sufficient to have inferred the components *A* and *B* separately, and to infer $A \supset B$, it is sufficient to add *A* temporarily to the stock of assumptions that have been made, then to infer *B*. In these rules, logical reasoning proceeds from the desired result to its deductive conditions. The reverse step is, then, to reason from an assumption or previously reached conclusion to its deductive consequences: to infer *A* from $A\&B$, to infer *B* from $A\&B$, and to infer *B* from $A \supset B$ and *A* together.

Gentzen's analysis of the structure of proofs in logic was a perfect success. He was able to show that the means for proving a logical theorem can be restricted to those that concern just the logical operations that appear in the theorem. Instead of logical axioms, there are just rules of inference, separately for each logical operation such as conjunction or implication, to the said effect. Logic on the whole is seen as a method for moving from given assumptions to a conclusion. The Fregean tradition, instead, presented logic as consisting of a basic stock of logical truths, namely the axioms of logic, together with two rules by which new logical truths can be proved from the axioms.

When Gentzen's logic is applied to axiomatic systems of mathematics, the axioms take their place among the assumptions from which logical proofs can start. It is commonly thought that Gentzen's analysis of the structure of proofs does not go through in such axiomatic extensions of pure logic. We try to show that this need not be so: the topic of this book is a method that treats axiomatic systems in a way analogous to Gentzen's natural deduction for pure logic, namely through the conversion of mathematical axioms into rules of inference, and with results analogous to those obtained in the proof analysis of pure logic.

1.2 Proof analysis: an introductory example

(a) **Natural deduction.** Gentzen's rules of natural deduction give an inductive definition of the notion of a **derivation tree**. Such a tree begins, i.e., has as leaves, formulas that are called **assumptions**. Each logical rule prescribes how a derivation tree (in brief, a derivation) of the conclusion of the rule is constructed from derivations of its premisses. The letter I indicates that a formula with a specific structure is concluded or **introduced**, and the letter E indicates that such a formula is, as one says, **eliminated**. For conjunction $A \& B$ and implication $A \supset B$, Gentzen gave the following rules:

Table 1.1 Gentzen's rules for conjunction and implication

$$\frac{A \quad B}{A \& B} \&I \qquad \frac{A \& B}{A} \&E \qquad \frac{A \& B}{B} \&E \qquad \frac{\overset{\displaystyle 1}{[A]} \atop \vdots \atop B}{A \supset B} \supset I,1 \qquad \frac{A \supset B \quad A}{B} \supset E$$

The rules, except for $\supset I$, are straightforward. In rule $\supset I$, a temporary assumption A is made, and a derivation of B from A can be turned into a derivation of $A \supset B$ by the rule. The square brackets indicate that the conclusion does not depend on the assumption A that has been **closed** or **discharged**. A **label**, usually a number, indicates which rule closes what assumptions.

Rules $\&I$ and $\supset E$ display one essential feature of Gentzen's work: they have two premisses so that derivation trees have binary branchings whenever these rules are applied. Each formula occurrence in a derivation tree determines a **subderivation** that lets us derive the formula, from precisely the assumptions it depends on. Often such subderivations can be rearranged combinatorially so that the same overall conclusion is obtained in a simpler way. Specifically, Gentzen's main result about natural deduction states that introductions followed by corresponding eliminations permit such rearrangements, with the effect that these steps of proof get removed from derivations. When no such simplifications are possible, all formulas in a derivation are parts or **subformulas** of the open assumptions or the conclusion. A brief expression is that **normal** derivations have the **subformula property**.

It is no exaggeration to say that the tree form of derivations that permits their transformation, in contrast to the earlier linear arrangement of Frege, Peano, Russell, and Hilbert and Bernays, was the key to all of Gentzen's central results: normalization in natural deduction, the corresponding method of cut elimination in sequent calculus, and the proof of the consistency of arithmetic.

Normalization consists in steps of **conversion** such as the following transformation of a part of a derivation:

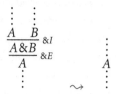

We shall need the normalizibility of logical derivations for the separation of logical and mathematical steps of proof. Gentzen's rules of natural deduction require some small changes presented in Chapter 2, before this separation can be made completely transparent.

(b) The theory of equality. We assume given a domain $\mathcal{D}$ of individuals, objects a, b, c ... of whatever sort, and a two-place relation $a = b$ in $\mathcal{D}$ with the following standard axioms:

Table 1.2 The axioms of an equality relation

EQ1	*Reflexivity:* $a = a$,
EQ2	*Symmetry:* $a = b \supset b = a$,
EQ3	*Transitivity:* $a = b \,\&\, b = c \supset a = c$.

These axioms can be added to a Frege–Hilbert-style axiomatization of logic. We shall instead first add them to natural deduction with the result that instances of the axioms can begin a derivation branch. Thus, when we ask whether a formula A is derivable from the collection of formulas Γ by the axioms of equality, arbitrary instances of the axioms can be added to Γ. We consider as an example a derivation of $d = a$ from the assumptions $a = b$, $c = b$, and $c = d$:

Table 1.3 A formal derivation in the axiomatic theory of equality

$$
\cfrac{
  \cfrac{a = c \,\&\, c = d \supset a = d \qquad
    \cfrac{
      \cfrac{a = b \,\&\, b = c \supset a = c \qquad
        \cfrac{
          \cfrac{c = b \supset b = c \quad c = b}{b = c}{\supset E}
          \quad a = b
        }{a = b \,\&\, b = c}{\&I}
      }{a = c}{\supset E}
      \qquad
      \cfrac{
        \cfrac{a = c \,\&\, c = d \supset a = d \quad \cfrac{c = d}{a = c \,\&\, c = d}{\&I}}{}{}
      }{}{}
    }{a = d}{\supset E}
  }{}{}
}{d = a}{}
$$

Each topformula in the derivation is either one of the atomic assumptions or an instance of an equality axiom. The derivation tree looks somewhat forbidding. The natural way to reason would be different, something like: *a is equal to b, b to c, c to d*, therefore *d is equal to a*. Here the principles are that equalities can be combined in chains and that equalities go both ways. The latter was applied to get the link *b equal to c* from *c equal to b*, and to get the conclusion *d equal to a* from *a equal to d*.

Logic in the derivation of $d = a$ from the assumptions $a = b$, $c = b$, and $c = d$ seems like some kind of a decoration necessitated by the use of logic in the writing of the axioms. We now want to say instead that $a = b$ gives at once $b = a$ and that two equalities $a = b$ and $b = c$ give at once $a = c$:

Table 1.4 Symmetry and transitivity as rules of inference

$$\frac{a = b}{b = a} \; Sym \qquad \frac{a = b \quad b = c}{a = c} \; Tr$$

Our example derivation becomes:

Table 1.5 A formal derivation by the rules for equality

$$\cfrac{a = b \quad \cfrac{\cfrac{c = b}{b = c} \; Sym}{a = c} \; Tr \quad c = d}{\cfrac{a = d}{d = a} \; Sym} \; Tr$$

This should be contrasted with the logical derivation of Table 1.3.

To get the full theory of equality, we must add reflexivity as a **zero-premiss** rule:

Table 1.6 The rule of reflexivity

$$\overline{a = a} \; Ref$$

Now formal derivations start from assumptions and instances of rule *Ref*.

What about the role of logic after the addition of mathematical axioms as rules? A premiss of an equality rule can be the conclusion of a logical rule and a conclusion of an equality rule a premiss in a logical rule. It should be clear that logic itself should not be 'creative' in the sense of making equalities derivable from given equalities used as assumptions, if they were not already derivable by just the equality rules. To show that there cannot be any such creative use of logic, Gentzen's normalization theorem comes to help. No introduction rule can have as conclusions premisses of a mathematical rule, because the latter do not have logical structure. Using a slight modification of Gentzen's elimination rules, the mathematical rules can be completely separated from the logical ones, so that in a normal derivation, the former are applied first, then the latter build up logical structure. Thus, if an equality is derivable from given equalities in natural deduction extended with the rules of equality, it is derivable by just the rules of equality. This separation of logic from mathematical axioms goes through for a large class of axiomatizations.

Assume there to be a derivation of the equality $a = c$ from given assumptions $a_1 = c_1, \ldots, a_n = c_n$ by the rules of equality. By what has been said, no logical rules need be used. Assume there to be a term b in the derivation that is neither a term in the conclusion $a = c$ nor a term in any of the assumptions. There is thus some instance of rule Tr that removes the **unknown** term b:

$$\frac{a = b \quad b = c}{a = c} \; Tr$$

If the premiss $a = b$ is a conclusion of rule Tr, we can permute up the instance of Tr that removes b, as follows:

$$\frac{\dfrac{a = d \quad d = b}{a = b} \; Tr \quad b = c}{a = c} \; Tr \qquad \rightsquigarrow \qquad \frac{a = d \quad \dfrac{d = b \quad b = c}{d = c} \; Tr}{a = c} \; Tr$$

A similar transformation applies if the second premiss $b = c$ has been derived by Tr. Thus, we may assume that neither premiss of the step of Tr that removes the term b has been derived by Tr. It can happen that both premisses have been derived by rule Sym. We then have a part of the derivation and its transformation:

$$\frac{\dfrac{b = a}{a = b} \; Sym \quad \dfrac{c = b}{b = c} \; Sym}{a = c} \; Tr \qquad \rightsquigarrow \qquad \frac{\dfrac{c = b \quad b = a}{c = a} \; Tr}{a = c} \; Sym$$

In the end, at least one premiss of the step of Tr that removes the term b has an instance of rule Ref as one premiss, as in

$$\frac{d = b \quad \dfrac{}{b = b} \; Ref}{d = b} \; Tr$$

Now the conclusion is equal to the other premiss, so the step of Tr can be deleted. Tracing up in the derivation the premiss $d = b$, the permutations can never lead to an instance of Tr that removes b and has an assumption as one premiss, because then b would be a term known from the assumption.

Thus, a derivation can be so transformed that it cannot have any unknown terms.

Consider next a derivation that has a 'cycle' or a 'loop', i.e., a branch with the same equality occurring twice:

The part between the two occurrences can be cut out. This part may use some equalities as assumptions that are not otherwise used in the derivation, but their deletion just improves the result: we would get the conclusion with fewer assumptions. When no loops are permitted, all derivations of an equality $a = c$ from the assumptions $a_1 = c_1, \ldots, a_n = c_n$ have an upper bound on size, here defined as the length of the longest derivation tree branch: the number of distinct terms is at most $2n + 2$; therefore the number of distinct equalities is at most $(2n + 2)^2$, an upper bound on height.

The above permutation argument could have been cut short as follows. If the equality to be derived is not an instance of *Ref*, that rule can be left out. If a premiss of *Sym* or *Tr* has been concluded by *Ref*, a loop is produced. Therefore all terms must appear in equalities that are assumptions. Such a simple argument does not usually work. The permutation argument, instead, illustrates a type of combinatorial reasoning that is characteristic of all that follows, beginning with the first real example, namely lattice theory in Chapter 4.

1.3 Outline

(a) **The four parts.** The book has four parts. The first is based on natural deduction in the sense that mathematical rule systems are formulated as extensions of the logical rules of natural deduction. These rules define a constructive system of logic in which existence proofs are effective and no classical case distinctions (A or $\neg A$) are made. All elimination rules are formulated in the manner of disjunction and existence elimination. As long as an axiom system contains no essential disjunctions, ones that cannot be converted into equivalent formulas without disjunctions, the logical rules can be permuted below the mathematical ones. Therefore parts of derivations by the latter rules can be separated from parts of derivations by the logical rules. The choice of classical or intuitionistic logic plays in this situation no role in the study of the derivations by such systems of mathematical rules.

With essentially disjunctive axioms, such as the linearity of an order relation, $a \leqslant b \vee b \leqslant a$, a classical sequent calculus formulation of logic permits the separation of logical and mathematical rules, in contrast to natural deduction. Sequent calculus was invented by Gentzen because he did not succeed in the proof analysis of classical logic formulated as a system of natural deduction. Part II of the book is based on sequent calculus in the sense that mathematical rule systems now extend the logical rules of sequent calculus.

We begin with axiomatic systems the axioms of which are universal, i.e., the axioms are quantifier-free formulas such as $a = b \,\&\, b = c \supset a = c$ in which a, b, and c are arbitrary parameters. Thus, such axioms could as well be written in the form $\forall x \forall y \forall z (x = y \,\&\, y = z \supset x = z)$. In Chapter 5 and in a general way in Part III, a much wider class of axioms is shown convertible to rules: those that are, in the terminology of category theory, geometric implications. Mathematical rules can now contain eigenvariables, which makes them behave like existential axioms, though without any visible logical structure.

Parts I–III build up gradually a method for an analysis of the structure of mathematical proofs. In each part, it is well defined to what kinds of axiomatic systems of mathematics the method can be applied. Part IV builds on all of the methods of the previous parts, but its focus is different. It occurred to the first author in 2003 that the method of proof analysis can be fruitfully applied to create systems of proof for modal logic and related non-classical logics: what is called the relational semantics of non-classical systems of logic, especially modal logic and its Kripke semantics, is formalized within the proof-theoretical calculi we use. The central new element, in comparison with Parts I–III, is the use of what are known as labelled logical calculi. Then, the properties that have been used previously on a semantical level can be represented by formulas that convert into rules just like the mathematical axioms treated in Parts I–III. It remains to be seen whether, in turn, the extension of purely logical proof systems in Part IV will find applications to more traditional mathematical structures.

(b) Summary of the individual chapters. The following is a list of the topics covered in the individual chapters, with an emphasis on new aspects that the method of proof analysis displays as well as on new results.

Chapter 2 presents natural deduction with general elimination rules, following von Plato (2001a). The extension of natural deduction by rules that correspond to axioms was indicated in Negri and von Plato (1998) and applied in Negri, von Plato, and Coquand (2004). Similar applications go in fact back to at least Prawitz (1971), and even Gentzen in the 1930s who converted arithmetical axioms into rules. The generalization of natural deduction is needed for separating parts of derivations by the rules of a mathematical theory from parts that use logical rules. Predicate logic with equality is formulated as such a theory, with two rules that act on atomic formulas. A proof-theoretical algorithm is given for deciding if an atomic formula is derivable from given atomic formulas used as assumptions.

Chapter 3 contains a discussion of axiomatic systems. It is very difficult to find organized treatments of the structure of an axiomatization. A rather detailed pattern for such structure is presented. Axioms with constructions are contrasted with ones that use additional relations to express what constructions do: say, a meet operation in lattice theory gives the constructed object $a \wedge b$ ('meet of a and b'), expressed relationally by the formula $M(a, b, c)$ ('the meet of a and b is c').

Chapter 4 treats order relations, lattice theory, and some equational theories. One of the most basic results of lattice theory is Skolem's 1920 positive solution to the word problem for freely generated lattices. This is the problem of how to decide whether a given atomic formula $a \leqslant b$ follows from a finite number of given atomic formulas used as assumptions. (Another name for this is the uniform word problem.) We give in Section 5.2 a permutation argument, from Negri and von Plato (2002), that establishes the subterm property (as explained above in Section 1.2) for derivations by the lattice rules. It is just a little over one page in length, does not assume any previous knowledge of anything, and gives, in our opinion, a direct glimpse into the 'combinatorial work' that is responsible for the positive solution. Section 4.3 treats the most basic structure in algebra, a domain of objects with a binary operation, i.e., a groupoid. It is shown that proof search in the word problem for groupoids can be restricted to terms known from the given atomic assumptions and from the conclusion. In Section 4.4, an alternative system is given for lattice theory, one that uses eigenvariables. The word problem is solved with almost no work. Secondly, the same method is applied to formalize the theory of strict partial order in the presence of equality.

In Chapter 5, existential axioms are converted into rules that extend the logical calculus of natural deduction. The class of axioms covered is a special case of the class of geometric implications. The special case consists in a restriction of the consequent of a geometric implication in such extensions, into one existential formula over a conjunction of atomic formulas instead of a disjunction of several such existential formulas. A dual of the class of geometric implications is defined, what we call co-geometric implications. Theories the axioms of which are such implications are called co-geometric theories. We have not found this class of theories treated systematically or even defined in the literature but were led to it by very natural considerations. Namely, existential formulas such as one that expresses the non-degeneracy of a strict partial order, $\exists x \exists y \, x < y$, contain an unnegated atomic formula and are geometric implications. An existential formula such as one that expresses the non-triviality of a domain with an equality relation, $\exists x \exists y \, \neg x = y$, contains a negated atomic formula and is a co-geometric implication. The latter is typical in many axiomatizations, such as projective and affine geometry in which the axiom of non-collinearity is co-geometric. Section 5.2 presents theories of equality and order as co-geometric and geometric theories, respectively. A relational axiomatization of lattice theory is given as a geometric theory in 5.3. It formalizes the system of lattice theory of Skolem (1920). It is shown that the existence axioms for meet and join are not needed in the proof search for an atomic formula from given atomic formulas used as assumptions. For the rest of the rules, it is shown that only those instances of the rules need be used in which all terms are known, which gives a bound on proof search.

When natural deduction is extended by rules that correspond to a relation such as a linear order $a \leqslant b$ with the linearity postulate $a \leqslant b \vee b \leqslant a$, there are two cases and it is not possible to convert the axiom into a rule in natural deduction style with the property that logical and mathematical parts of derivations always remain separate. For this reason, the logical system of proof used is changed into a sequent calculus in Part III. Chapter 6 begins with an introduction to sequent calculus. Its rules are motivated directly from those of natural deduction with general elimination rules. Next the extension of sequent calculus by mathematical rules is presented. The restriction in natural deduction to formulas that do not contain essential disjunctions can be dropped, so that first a scheme is presented for converting any universal formula into a rule, or several rules. As with natural deduction, predicate logic with equality is given as a first example of the extension of sequent calculus. It is shown through proof analysis that predicate logic with equality is conservative over predicate logic. Derivations without equalities in the endsequent are so transformed that possible applications of rules for equality get eliminated from them. In Section 6.4, a generalization of Herbrand's theorem to universal theories is given.

Chapter 7 presents linear order as an extension of sequent calculus. A proof-theoretical algorithm is given in Section 7.1 for ordering linearly any partially ordered set. The possibility of such ordering is known as Szpilrajn's theorem. We show that if $\mathcal{D}$ is a set with a weak partial order $a \leqslant b$, Δ a finite set of atomic formulas, and c and d two terms in Δ such that neither $c \leqslant d$ nor $d \leqslant c$ is included in the partial order in $\mathcal{D}$, then either $c \leqslant d$ or $d \leqslant c$ can be consistently added to Δ. Section 7.2 gives a proof of the subterm property for derivations by the rules of linear order. Section 7.3 extends this proof into a corresponding result for linear lattices, i.e., lattices in which the ordering relation is linear.

Chapter 8 contains a study of the extension of sequent calculi by rules that correspond to geometric implications, without the restriction that had to be made in natural deduction in Chapter 5. In Section 8.3, a proof-theoretical approach to a fundamental result of category theory, namely Barr's theorem, is given. The theorem states that if a geometric implication is derivable in a geometric theory by the use of classical logic, then it is already derivable by the use of intuitionistic logic. Our proof of this result consists in noting that in a suitable rule system for geometric theories, a derivation of a geometric implication is necessarily one in which the intuitionistic restrictions on the rules of implication and universal quantification are met.

The system of natural deduction of Part I is intuitionistic. In sequent calculus, both classical and intuitionistic versions of the logical calculus can be extended by rules. Chapter 9 studies classical and intuitionistic axiomatizations, such as a classical equality relation and an intuitionistic apartness relation. It is noted that derivations by the corresponding rules are duals, in the sense that if the intuitionistic basic notions and axioms are converted into a rule system that acts on the antecedent part of sequents, a mirror-image system of rules acts on the right part of sequents and corresponds to classical basic notions and axioms. Examples beyond equality and apartness contain classical and intuitionistic projective and affine geometry. The combinatorial challenge of proof analysis of derivations by the rules of an intuitionistic or a classical theory is exactly the same. An essential difference between intuitionistic and classical theories can be met at the stage of conversion of axioms into rules: the conjunctive normal form that we use in such conversion may not be intuitionistically equivalent to an axiom, whereas classically every universal formula has an equivalent in conjunctive normal form. Secondly, derivations by the logical rules are different. The latter, however, has no effect on the mathematical parts of derivations that are combinatorially duals of each other. Chapter 9 presents as an application of the duality a generalization of Herbrand's theorem to geometric theories, and another one to co-geometric theories.

In Chapter 10, plane projective and affine geometries are converted into proof systems. Both contain an existential axiom, that of non-collinearity, by which there exist at least three non-collinear points. At the same time, there are axioms that contain essential disjunctions. To make the combinatorial analysis of derivations more manageable, a multiple-conclusion calculus in the style of natural deduction is used. It is shown that the derivability of a finite number of atomic cases from a finite number of atomic assumptions is decidable in both geometries. The existential axiom turns out to be conservative over the rest of the axioms for the said derivability problem.

The methods of proof analysis are applied to modal logic in Chapter 11. The main idea of the approach is to make the well-known Kripke semantics of modal logic a part of the formalism, through the use of labels that correspond to possible worlds. A sequent calculus for basic normal modal logic is given that has all the standard properties, namely that the structural rules are admissible. The calculus is furthermore suited to root-first proof search because its rules are invertible, with height of derivation preserved. The extensions of basic modal logic into various modal calculi are obtained by adding rules to the calculus that correspond to frame properties. Most such properties are expressible as universal axioms or geometric implications. Therefore the structural properties of the extensions can be guaranteed once and for all, similarly to any extensions of sequent calculi by mathematical rules. In Section 11.5, termination of proof search for most modal calculi is shown through a proof of the subterm property for minimal derivations. In Section 11.6, modal undefinability results, usually obtained through model-theoretic arguments, follow from simple analyses of derivations. In the final section of Chapter 11, a detailed proof of the completeness of a calculus for basic modal logic is given, in the same spirit as Kripke's original proof of 1963, but with the added formalism of labels.

Chapter 12 begins with the addition of the quantifiers to basic modal logic. Then provability logic is treated, with cut elimination proved with no compromises. In the final section, labelled sequent calculi are given for relevance and other substructural logics and intermediate logics. The idea is always to use the frame properties of the relational semantics for these logics, through the conversion of these properties into corresponding rules by which the basic calculus is extended.

PART I

Proof systems based on natural deduction

2 | Rules of proof: natural deduction

This chapter gives, first, the calculus of natural deduction, together with its basic structural properties such as the normalization of derivations and the subformula property of normal derivations. Next, the calculus is extended by mathematical rules, and it is shown that normalization works also in such extensions. The theory of equality is treated in detail, as a first example. Finally, predicate logic with an equality relation is studied. It is presented as an extension of predicate logic without equality, and therefore normalization of derivations applies. The question of the derivability of an atomic formula from given atomic formulas, i.e., the word problem for predicate logic with equality, is solved by a proof-theoretical algorithm.

2.1 Natural deduction with general elimination rules

Gentzen's rules of natural deduction for intuitionistic logic have proved to be remarkably stable. There has been variation in the way the closing of assumptions is handled. In 1984, Peter Schroeder-Heister changed the rule of conjunction elimination so that it admitted an arbitrary consequence similarly to the disjunction elimination rule. We shall do the same for the rest of the elimination rules and prove normalization for natural deduction with general elimination rules.

Natural deduction is based on the idea that proving begins in practice with the making of assumptions from which consequences are then drawn. Thus, the first rule of natural deduction is that any formula A can be assumed. Formally, by writing

A

we construct the simplest possible derivation tree, that of the conclusion of A from the assumption A.

(a) **Introduction rules as determined by the BHK-conditions.** As explained by Gentzen, the introduction rules formalize natural conditions on direct proofs of propositions of the different logical forms. These are often referred to as the BHK-conditions (for Brouwer, Heyting, and Kolmogorov) by which $A \& B$ is proved directly by proving A and B separately, $A \lor B$ by proving one of A and B, and $A \supset B$ by proving B from A. The rules are:

Table 2.1 Gentzen's introduction rules

$$\frac{A \quad B}{A \& B} \, \&I \qquad \frac{A}{A \lor B} \, \lor I_1 \qquad \frac{B}{A \lor B} \, \lor I_2 \qquad \frac{\overset{\displaystyle 1}{[A]}}{\overset{\displaystyle \vdots}{\underset{A \supset B}{B}}} \, \supset I, 1$$

In rule $\supset I$, the formula A is assumed temporarily in order to derive B. The notation $[A]$ indicates that the assumption is **closed** or **discharged** at the inference. It is possible that the assumption has been used several times to infer B, or even 0 times. Not all instances of the assumption need be closed, even if this would usually be the case. The number next to the rule and on top of formulas is a **discharge label**.

For the quantifiers, we have:

Table 2.2 Introduction
rules for the quantifiers

$$\frac{A(y/x)}{\forall x A} \, \forall I \qquad \frac{A(t/x)}{\exists x A} \, \exists I$$

The notation $A(y/x)$ stands for the **substitution** of free occurrences of the variable x in formula A by the variable y ('y for x'). Rule $\forall I$ has the standard variable restriction: the **eigenvariable** y must not occur free in any assumptions the premiss $A(y/x)$ depends on.

(b) Inversion principle: determination of elimination rules. Gentzen noticed that the elimination rules of natural deduction (E-rules) somehow repeat what was already contained in derivations with corresponding introduction rules (I-rules), and speculated that it should be possible to actually determine E-rules from I-rules. The idea is captured by the following:

Inversion principle. *Whatever follows from the direct conditions for introducing a formula, must follow from that formula.*

The principle determines the following **general elimination rules**, with a slight proviso on implication:

Table 2.3 General elimination rules for the connectives

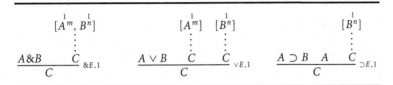

The number of instances of a formula that are closed is indicated by an exponent. In general, any numbers $m, n \geqslant 0$ of open assumptions A and B can be chosen to be closed. Some instances of the same formula may become closed while others remain open. If $m = 0$ or $n = 0$, there is a **vacuous** discharge; if $m > 1$ or $n > 1$, there is a **multiple** discharge. Otherwise a discharge is **simple**. Each instance of a rule must have a fresh label.

The standard elimination rules of natural deduction follow by setting, in turn, $C \equiv A$ or $C \equiv B$ in $\&E$, and $C \equiv B$ in $\supset E$. The derivations of the minor premiss C become degenerate.

A direct proof of $A \supset B$ consists in a derivation of B from the assumption A. Thus, our inversion principle dictates that C follows from $A \supset B$ if C follows from the existence of a derivation of B from A. First-order logic cannot express this situation, so rule $\supset E$ just shows how arbitrary consequences of B reduce to arbitrary consequences of A under the major premiss $A \supset B$. Schroeder-Heister, instead, used a higher-order rule, and so does type theory.

The propositional part of intuitionistic natural deduction is completed by adding an elimination rule for $\bot$ and by defining negation and equivalence:

Table 2.4 Falsity elimination, negation, and equivalence

$$\frac{\bot}{C} \bot E \qquad \neg A \equiv A \supset \bot \qquad A \supset\subset B \equiv (A \supset B)\&(B \supset A).$$

Finally, we have the following table:

Table 2.5 Elimination rules for the quantifiers

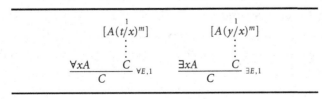

In rule $\forall E$, t is any term, i.e., a constant or a variable. In rule $\exists E$, the eigenvariable y must occur free neither in the conclusion nor in any other open assumption the minor premiss C depends on except $A(y/x)$.

The standard elimination rule for the universal quantifier follows by setting $C \equiv A(t/x)$. The derivation of the minor premiss C is then degenerate, similarly to the propositional case. When these degenerate derivations are left unwritten in our rules, we get:

Table 2.6 Standard elimination rules of natural deduction

$$\frac{A\&B}{A}\;{\scriptstyle \&E} \qquad \frac{A\&B}{B}\;{\scriptstyle \&E} \qquad \frac{A \supset B \quad A}{B}\;{\scriptstyle \supset E} \qquad \frac{\forall x A}{A(t/x)}\;{\scriptstyle \forall E}$$

(c) **Discharge principle: definition of derivations.** 'Compulsory discharge' dictates that one must discharge if one can. But look at

$$\frac{\dfrac{\overset{1}{[A]}}{B \supset A}\;{\scriptstyle \supset I}}{A \supset (B \supset A)}\;{\scriptstyle \supset I,1}$$

The upper rule is one with a vacuous discharge. Assumption A is closed at the second step. If it happened that B is identical to A, compulsory discharge would require a discharge of A at the first step, so something that looked like a syntactically correct derivation under the 'compulsory' idea turned out not to be so. We adopt instead the following:

Discharge principle. *Each rule instance must have a fresh discharge label.*

The example inference is now written

$$\frac{\dfrac{\overset{2}{[A]}}{B \supset A}\;{\scriptstyle \supset I,1}}{A \supset (B \supset A)}\;{\scriptstyle \supset I,2}$$

If we have A in place of B, the derivation remains formally correct.

We can now give a formal definition of the notion of a **derivation of formula A from the open assumptions Γ**. The open assumptions are counted with multiplicity, so they are multisets of formulas. The base case of a derivation is the derivation of a formula A from the open assumption A:

A

Now the rest are defined inductively according to the last rule applied, straightforward for rules that do not change the assumptions, and exemplified by $\&E$ for the rest:

Given derivations of $A\&B$ from Γ and of C from A^m, B^n, Δ,

$$
\frac{\begin{array}{cc} \Gamma & [\overset{1}{A^m}, \overset{1}{B^n}], \Delta \\ \vdots & \vdots \\ A\&B & C \end{array}}{C}\ \&E,1
$$

is a derivation of C from Γ, Δ.

The full definition is given in *Structural Proof Theory*, pp. 167–170. We observe that the putting together of two derivations is justified by the definition:

Composition of derivations. *Given derivations of A from Γ and of C from A, Δ with no clash on labels and eigenvariables, they can be put together into a derivation of C from Γ, Δ.*

The closing of assumptions needs to be treated explicitly for composition to produce a correct derivation. If two derivations to be composed use the same eigenvariable symbols, these can be changed before the composition. Similarly, in any given derivation, we can assume that rules with eigenvariables have distinct eigenvariables and that each eigenvariable occurs only in the part of the derivation above its associated rule of inference.

(d) Derivable and admissible rules. The **derivability** of a rule in a given system of rules requires that the conclusion of the rule be derivable from its premisses. There is an important related notion that is weaker, namely the **admissibility** of a rule in a given system of rules, defined by: whenever the premisses of the rule are derivable in the system, the conclusion also is. Proofs of admissibility of a rule consider the ways in which the premisses can have been derived, i.e., admissibility is proved by an induction on the last rules applied in the derivations of the premisses.

(e) Classical propositional logic. The system of introduction and elimination rules gives what is called **intuitionistic logic**. The reason for this historically established nomenclature is that there is no rule that would correspond to a principle of indirect proof. Natural deduction for **classical**

propositional logic can be obtained by adding to the rules of intuitionistic propositional logic a **rule of excluded middle**:

$$\frac{\overset{1}{[A]} \quad \overset{1}{[\neg A]}}{\underset{\vdots}{C} \quad \underset{\vdots}{C}}{C} \; Em,1$$

Both A and $\neg A$ are closed at the inference. The law of excluded middle, $A \vee \neg A$, is derivable by the rule:

$$\frac{\dfrac{\overset{1}{[A]}}{A \vee \neg A} \vee I_1 \quad \dfrac{\overset{1}{[\neg A]}}{A \vee \neg A} \vee I_2}{A \vee \neg A} \; Em,1$$

The rule of excluded middle is a generalization of the **rule of indirect proof** (*reductio ad absurdum*):

$$\frac{\overset{1}{[\neg A]}}{\underset{\vdots}{\bot}}{A} \; Raa,1$$

This rule can be derived as follows from rule *EM*:

$$\frac{\overset{1}{[A]} \quad \dfrac{\overset{1}{[\neg A]}}{\dfrac{\underset{\vdots}{\bot}}{A} \bot E}}{A} \; Em,1$$

The difference between a genuine indirect inference and an inference to a negative proposition is not always appreciated. In the latter, there is a derivation of $\bot$ from an assumption A, by which $\neg A$ can be inferred. This step is not, however, made by rule *Raa*, but by a special case of rule $\supset I$. Often proofs of irrationality of a real number, say $\sqrt{2}$, are given as examples of indirect inferences, which they are not, because the property to be proved is negative.

A system of classical propositional logic is obtained also when the rule of excluded middle is restricted to atomic formulas (*cf. Structural Proof Theory*, section 8.6). The rule is shown admissible for arbitrary propositional formulas. Its admissibility for arbitrary formulas means that its use can be replaced by uses of the rule on atomic formulas. The reduction of the rule of excluded middle to atoms does not go through when the quantifier rules are added.

2.2 Normalization of derivations

(a) Convertibility. The standard normalization procedure of Gentzen and Prawitz consists in the removal of consecutive introduction–elimination pairs. Such pairs are known as **detours**, or *Umwege* in Gentzen's terminology, and their eliminations **detour conversions**. Later Prawitz considered **permutation convertibilities**, instances of $\vee E$ or $\exists E$ that have as a conclusion a major premiss of an E-rule, and **simplification convertibilities**.

Definition 2.1. *An E-rule with a major premiss derived by an I-rule is a* **detour convertibility**.

A detour convertibility on $A\&B$ and the result of the conversion are, with obvious labels left unwritten,

$$
\frac{\dfrac{A \quad B}{A\&B}\;{\&}I \qquad \overset{\textstyle [A^m,\,B^n]}{\underset{\textstyle C}{\vdots}}}{C}\;{\&}E
\qquad\qquad
\underset{C}{\overset{A\;\overset{m\times}{\cdots}\;A\quad B\;\overset{n\times}{\cdots}\;B}{\vdots}}
$$

A detour convertibility on disjunction is quite similar. A detour convertibility on $A \supset B$ and the result of the conversion are

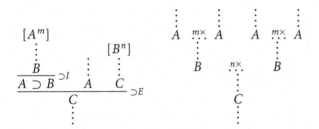

There is no I-rule for $\bot$, so no detour convertibility either.

Definition 2.2. *An E-rule with a major premiss derived by an E-rule is a* **permutation convertibility**.

The novelty of general elimination rules is that permutation conversions apply to all cases in which a major premiss of an E-rule has been derived. With six E-rules, this gives 36 convertibilities of which we show a couple:

A permutation convertibility on major premiss $C\&D$ derived by $\&E$ on $A\&B$ and its conversion are

$$
\cfrac{\cfrac{A\&B \qquad C\&D}{C\&D}\ {}_{\&E} \qquad \overset{\displaystyle[C^k,D^l]}{\underset{\displaystyle E}{\vdots}}}{E}\ {}_{\&E}
\qquad\qquad
\cfrac{A\&B \qquad \cfrac{\overset{\displaystyle[A^m,B^n]}{\underset{\displaystyle C\&D}{\vdots}} \qquad \overset{\displaystyle[C^k,D^l]}{\underset{\displaystyle E}{\vdots}}}{E}\ {}_{\&E}}{E}\ {}_{\&E}
$$

with assumptions $[A^m, B^n]$ above $A\&B$.

A permutation convertibility on major premiss $C \supset D$ derived by $\vee E$ on $A \vee B$ obtains whenever a derivation has the part

$$
\cfrac{\cfrac{A \vee B \quad \overset{[A^m]}{\underset{C\supset D}{\vdots}} \quad \overset{[B^n]}{\underset{C\supset D}{\vdots}}}{C \supset D}\ {}_{\vee E} \qquad C \qquad \overset{[D^l]}{\underset{E}{\vdots}}}{E}\ {}_{\supset E}
$$

After the permutation conversion the part is

$$
\cfrac{A \vee B \qquad \cfrac{\overset{[A^m]}{C\supset D} \quad \overset{[D^l]}{\underset{E}{\vdots}}\ C \ \ E}{E}\ {}_{\supset E} \qquad \cfrac{\overset{[B^n]}{C\supset D} \quad \overset{[D^l]}{\underset{E}{\vdots}}\ C\ \ E}{E}\ {}_{\supset E}}{E}\ {}_{\vee E}
$$

Finally, we have permutation convertibilities in which the conversion formula is $\bot$ derived by $\bot E$. Since $\bot E$ has only a major premiss, a permutation conversion just removes one of these instances:

$$
\cfrac{\cfrac{\bot}{\bot}\ {}_{\bot E}}{C}\ {}_{\bot E} \qquad \rightsquigarrow \qquad \cfrac{\bot}{C}\ {}_{\bot E}
$$

Definition 2.3. A **simplification convertibility** *in a derivation is an instance of an E-rule with no discharged assumptions, or an instance of $\vee E$ with no discharges of at least one disjunct.*

As with permutation conversions, simplification conversions also apply to all *E*-rules when general elimination rules are used. A simplification convertibility can prevent the normalization of a derivation, as is shown by the following:

$$\frac{\dfrac{\overset{1}{[A]}}{A \supset A}\supset\!I,1 \quad \dfrac{\overset{2}{[B]}}{B \supset B}\supset\!I,2}{\dfrac{(A \supset A)\&(B \supset B)}{C \supset C}}\&I \quad \dfrac{\overset{3}{[C]}}{C \supset C}\supset\!I,3 \ \&E$$

There is a detour convertibility but the pieces of derivation do not fit together in the right way to remove it. Instead, a simplification conversion will remove the detour convertibility:

$$\frac{\overset{3}{[C]}}{C \supset C}\supset\!I,3$$

Note that the notion of simplification convertibility is somewhat different from those of detour and permutation convertibilities. In the latter two, the major premiss of an *E*-rule is derived, whereas it can be an assumption in a simplification convertibility.

(b) Normal derivations. In Gentzen's natural deduction, the possible convertibilities are, first, detour convertibilities on all the connectives and quantifiers. Secondly, there are permutation convertibilities on disjunction and existence, and simplification convertibilities likewise on disjunction and existence. A derivation can be defined to be normal in Gentzen's natural deduction when it has no such convertibilities. The use of general elimination rules leads to a remarkable simplification of the notion of normal derivability:

Definition 2.4. *A derivation is* **normal** *if all major premisses of elimination rules are assumptions.*

As already indicated, a normal derivation can still contain simplification convertibilities.

The definition of normal derivations extends also to the system of classical propositional logic of Section 2.1(e). The rule of excluded middle can be permuted down relative to the introduction and elimination rules. A derivation is normal if it has normal intuitionistic subderivations followed by rules of excluded middle. It further follows that for a normal derivation, the atoms in the rules of excluded middle are subformulas of open assumptions or of the conclusion. Thus, each formula in a normal derivation is a subformula of an open assumption or the conclusion, or a negation of an atomic subformula.

(c) **The subformula structure.** Due to the form of the general E-rules we consider subformula structure along **threads** in a derivation (a term suggested to us by Dag Prawitz), instead of branches of the derivation tree as would be the case for Gentzen's elimination rules in the $\vee, \exists$-free fragment. These threads are constructed starting with the endformula of a derivation:

1. For I-rules, the threads are

$$\frac{A}{A\&B} \quad \frac{B}{A\&B} \quad \frac{A}{A\vee B} \quad \frac{B}{A\vee B} \quad \frac{B}{A\supset B} \quad \frac{A(y/x)}{\forall xA} \quad \frac{A(t/x)}{\exists xA}$$

2. For E-rules, the thread continues up from the minor premiss C, with two threads produced for $\vee E$:

$$\frac{C}{C}$$

3. If the last formula is an open assumption A or an assumption A closed by $\supset I$, the thread ends with topformula A.

4. If the last formula is an assumption A or B closed by $\&E$ or $\vee E$, the construction of the thread continues from the major premiss $A\&B$ or $A\vee B$:

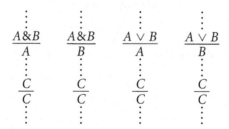

The construction is similar for rules $\forall E$ and $\exists E$.

5. With $\bot E$, there is no minor premiss, so the construction continues directly from the major premiss $\bot$:

6. If the last formula is an assumption B closed by $\supset E$, the construction continues with the major premiss $A \supset B$. A **new thread** begins with the minor premiss A as endformula:

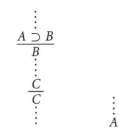

Note that the construction of threads will not reach the parts of derivation that can be deleted in a simplification convertibility.

We can depict threads as follows, with a semicolon separating the ith major premiss of an E-rule A_{h_i} from its components C_{h_i} closed by the elimination:

$$(A_1, \ldots, A_{h_1}; C_{h_1}, \ldots, A_{h_2}; C_{h_2}, \ldots, A_{h_i}; C_{h_i}, \ldots, A)$$

(d) The normalization of derivations. The **height** of a major premiss A_{h_i} in a thread is measured as follows. Let h_1 be the number of steps from the topformula to a first major premiss A_{h_1} and h_i the number of steps from the temporary assumption of the preceding major premiss $A_{h_{i-1}}$ to A_{h_i}. The height of A_{h_i} in the thread is the sum $h_1 + \cdots + h_i$.

From the construction of threads it is immediately evident that each formula in a derivation is in at least one thread. A thread is **normal** if it is a thread of a normal derivation. The height of each major premiss in normal threads is equal to zero. It is easily seen that the converse also holds. The formulas in a thread divide into an 'E-part' of nested sequences of major premisses, each a subformula of the preceding formula, and an 'I-part' in which formulas start building up in the other direction through introduction rules. Each formula in a normal thread is a subformula of the endformula or of an open assumption. (For a proof, not difficult, see *Structural Proof Theory*, p. 197.)

The proof of normalization uses what is called a **multiset ordering**. In the proof, conversion formulas in threads in a derivation are collected into multisets according to their length. The basic property of multiset ordering is its well foundedness, which is used in the following way. If through a detour conversion a conversion formula of length n is removed and replaced by any number of conversion formulas of a strictly lesser length, the multiset ordering of the conversion formulas is reduced. Well-foundedness means that in the end, no convertibilities remain.

Lemma 2.5. *A permutation conversion on major premiss A diminishes its height by one and leaves all other heights unaffected.*

Given a derivation, consider its conversion formulas in each thread, ordered by length into multisets.

Lemma 2.6. *Detour conversions on & and $\vee$ reduce the multiset ordering of conversion formulas in threads affected by the conversion.*

Note that permutation conversions do not create any new conversion formulas and therefore do not affect the multiset ordering. They can change a permutation convertibility into a detour convertibility. If this happens with implication, a new thread with the minor premiss as endformula is constructed.

The construction of threads is essential in Lemma 2.6. It is seen from the detour conversion scheme for & that parts of the derivation get multiplied. These parts can contain conversion formulas of any length, so the multiset of conversion formulas for the whole derivation is not necessarily reduced. For threads, instead, it is reduced.

For each case of detour convertibility on implication, we consider the derivation in three parts:

1. The derivations of the first minor premiss A.
2. The derivations of the second minor premiss C.
3. The derivations of the major premiss $A \supset B$ and the continuation from the conclusion C.

The idea of normalization is to postpone detour conversions on implication. Assume therefore steps of normalization other than detour on implication to have been made. No conversion can create new major premisses of E-rules. Therefore only a bounded number of detour convertibilities on implication can be met in such normalization, and both of the minor premisses in an uppermost instance of $\supset E$ have a normal derivation. Possible detour conversion on such instances can create new convertibilities, but they are on strictly shorter formulas. An iteration of this procedure will eventually lead to a downmost detour convertibility on implication to be eliminated, and we have:

Theorem 2.7. *Natural deduction with general elimination rules is normalizing.*

(e) Strong normalization. The condition of **strong** normalization of a system of natural deduction means that conversions terminate, irrespective of the order in which they are made. Joachimski and Matthes (2003) proved directly strong normalization for natural deduction with general elimination rules. Their proof uses a system of term assignment. The above proof of normalization is 'almost strong', in that the only restriction on conversions concerns the postponing of normalization at detour convertibilities on implication. It would be interesting to find a simple argument, based on the conversion schemes and their combinatorial behaviour, by which this restriction can be lifted.

2.3 From axioms to rules of proof

(a) Mathematical rules. We shall extend our calculus of intuitionistic natural deduction by suitably formulated **mathematical rules**. Assume a mathematical theory the axioms of which are all universal formulas in predicate calculus. Each such formula can be represented as a conjunction of implications that have the form

$$P_1 \& \ldots \& P_m \supset Q_1 \vee \ldots \vee Q_n \tag{1}$$

with the P_i, Q_j atomic formulas. It turns out that implications of this form can be converted into rules that permit a separation of the mathematical and logical part of the derivation if $n \leqslant 1$.

Definition 2.8. *The **Harrop** formulas of propositional logic are:*

(i) *Atomic formulas and $\perp$ are Harrop.*
(ii) *If A and B are Harrop, also $A \& B$ is Harrop.*
(iii) *If B is Harrop, also $A \supset B$ is Harrop.*

The idea is that there are no cases (disjunctions) among Harrop formulas, nor any cases 'hidden' inside implications, such as in $A \supset B \vee C$. On the other hand, some disjunctions are inessential, in that there are equivalent disjunction-free formulas, such as $A \vee B \supset C$ and $(A \supset C) \& (B \supset C)$.

With $n = 1$ in (1), we have

$$P_1 \& \ldots \& P_m \supset Q \tag{2}$$

We convert axioms of this form into rules of the form

$$\frac{P_1 \quad \ldots \quad P_m}{Q} \tag{3}$$

As one limiting case we have $m = 0$ and $n = 1$, and a **zero-premiss** rule:

$$\frac{}{Q} \tag{4}$$

As another limiting case we have $n = 0$, as best seen from (1). The false formula $\perp$ is put in the place of the consequent and (1) becomes the negation

$$\neg(P_1 \& \ldots \& P_m) \tag{5}$$

The corresponding rule has $\perp$ in place of the formula Q of (3):

$$\frac{P_1 \quad \cdots \quad P_m}{\perp} \tag{6}$$

Derivations by mathematical rules are finitely branching trees with atoms at the nodes. A finite number of axioms of the form (2) gives as many rules of the form (3). An extension of the system of natural deduction **NI** by such rules is denoted by **NI***.

Lemma 2.9. *In a derivation in* **NI***, *instances of logical rules can be permuted to apply after the mathematical rules.*

Proof. Consider a premiss of a mathematical rule R, say P_1, derived by a logical rule. The I-rules give as conclusions compound formulas, so the only possible rules are E-rules. If the major premiss is $A\&B$ and minor premiss P_1 in rule $\&E$ followed by rule R, we have a part of the derivation:

$$
\cfrac{A\&B \quad \cfrac{[A,B] \\ \vdots \\ P_1}{}}{\cfrac{P_1}{ } \&E \quad P_2 \quad \cdots \quad P_n}{Q} R
$$

The permuted part of the derivation is

$$
\cfrac{A\&B \quad \cfrac{\cfrac{[A,B] \\ \vdots \\ P_1 \quad P_2 \quad \cdots \quad P_n}{Q} R}{Q}}{Q} \&E
$$

Note that with rule $\exists E$, we can assume that its eigenvariable occurs only above the rule in the derivation, therefore not in the derivations of the premisses $P_2, \ldots, P_n$. No violation of the eigenvariable condition is thus produced when these derivations are permuted to above rule $\exists E$. The transformation is similar for the rest of the E-rules except $\bot E$. In a purely logical derivation, normalization brings with itself that also the premiss of $\bot E$ is an assumption, but now it can be concluded by a mathematical rule of the form given in (6) above. In this case we have a part of the derivation and its transformation:

$$\dfrac{\dfrac{\dfrac{Q_1 \ldots Q_m}{\bot} \, {\scriptstyle R1}}{P_1} \, {\scriptstyle \bot E} \quad \ldots P_n}{Q} \, {\scriptstyle R2} \qquad \dfrac{\dfrac{\dfrac{Q_1 \ldots Q_m}{\bot} \, {\scriptstyle \bot E}}{Q} \, {\scriptstyle \bot E}}{}$$

If anywhere in a derivation a premiss of a mathematical rule has been concluded by $\bot E$, the mathematical rule can be deleted. The repetition of permutations such as for $\&E$ brings all E-rules below mathematical rules.   QED.

It can be seen from the two schemes in the proof that permutation of logical rules to below mathematical rules does not affect the heights of derivation of major premisses of E-rules. Therefore the normalization theorem extends to derivations in **NI***.

Theorem 2.10. *Derivations in* **NI*** *convert to a form in which major premisses of E-rules are assumptions except for $\bot$ that can have been derived by a mathematical rule.*

Proof. The proof is an extension of the proof of normalization for **NI**. The major premiss of an E-rule is never concluded by a mathematical rule except for $\bot E$. Therefore the possible convertibilities are the same as in a purely logical derivation, and mathematical rules do not interfere in any way with the conversion process.   QED.

If intuitionistic natural deduction **NI** is changed into **minimal logic** through the leaving out of rule $\bot E$, no qualification is needed in Theorem 2.10.

(b) **The subterm property.** We can now define a property that will give the decidability of the derivability of an atomic formula from a finite number of atomic formulas by the rules of a theory:

Definition 2.11. *A derivation of an atom P from atomic assumptions Γ in a system of mathematical rules has the* **subterm property** *if all terms in the derivation are terms in Γ, P. The derivation is* **loop free** *if no atom appears twice in a derivation branch.*

Terms shall be called **known** terms if they appear in Γ, P, and otherwise **new** terms.

Definition 2.12. *A derivation of an atom P from atomic assumptions Γ in a system of mathematical rules is **minimal** if it is loop free and atoms that can be concluded by a zero-premiss rule are concluded by such a rule.*

(c) Complexity of derivations. Mathematical rules of the form (3) can be converted into **program clauses** that have the form:

$$P_1, \ldots P_m \to Q$$

These program clauses can be combined through a rule of **composition**, as in:

$$\frac{P_1, \ldots P_m \to Q_1 \quad Q_1, \ldots Q_n \to R}{P_1, \ldots P_m, Q_2, \ldots Q_n \to R}\ Comp$$

Composition obviously corresponds to the application of two rules in succession, so the composition formula is the conclusion of the first rule and a premiss of the second. The general derivability problem that we consider within systems that extend natural deduction can be expressed in terms of clauses. Let the assumptions be the atomic formulas $S_1, \ldots S_k$ and the atom to be derived T. Let the clauses that correspond to the axioms be

$$P_1, \ldots P_m \to Q, \quad \ldots \quad Q_1, \ldots Q_n \to R. \tag{7}$$

If minimal derivations have the subterm property, the derivability problem becomes: Is the clause $S_1, \ldots S_k \to T$ derivable by *Comp* from those instances of (7) that contain only terms known from the clause to be derived? There is only a bounded number of distinct clauses with known terms, and all of the clauses have at most one formula at the right of the arrow. It is one of the basic results of logic programming that this derivability problem can be decided in a number of steps that is polynomial in the number of data, i.e., the size of the clause to be derived.

2.4 The theory of equality

(a) The rules of equality. The axioms of an equality relation convert into the following system of rules:

$$\frac{}{a = a}\ Ref \qquad \frac{a = b}{b = a}\ Sym \qquad \frac{a = b \quad b = c}{a = c}\ Tr$$

We shall go through in detail the proof of the subterm property for the theory of equality, already described in the Introduction.

Lemma 2.13. Subterm property for the theory of equality. *Minimal derivations of atoms from atomic assumptions in the theory of equality have the subterm property.*

Proof. Consider a minimal derivation, and let b be an uppermost new term, i.e., a new term such that no other new term has been removed from the derivation above the last appearance of b. We then have a subderivation that ends with a last occurrence of b:

$$\frac{a = b \quad b = c}{a = c} \; Tr$$

If the left premiss is derived by Tr, we have, for some d, the derivation

$$\frac{\dfrac{a = d \quad d = b}{a = b} \; Tr \quad b = c}{a = c} \; Tr$$

The step that removes the new term b is permuted up:

$$\frac{a = d \quad \dfrac{d = b \quad b = c}{d = c} \; Tr}{a = c} \; Tr$$

If the right premiss of the original Tr has been derived by Tr, a similar permutation is made. Permuting up the instance of Tr removing b, it is possible that both premisses become derived by Sym:

$$\frac{\dfrac{b = d}{d = b} \; Sym \quad \dfrac{e = b}{b = e} \; Sym}{d = e} \; Tr$$

This is transformed into

$$\frac{\dfrac{e = b \quad b = d}{e = d} \; Tr}{d = e} \; Sym$$

In the end, the instance of Tr removing the new term b has as one premiss a topformula. It cannot be an assumption because b was supposed to be a new term. If it is an instance of *Ref*, the conclusion is equal to the other premiss, against the assumption of a minimal derivation. QED.

Theorem 2.14. *The derivability of an atom from a given number of atoms by the rules of the theory of equality is decidable.*

Proof. By the subformula property, proof search can be restricted to the known terms. If there are n such terms, there are at most n^2 distinct atoms with these terms. The number of loop-free derivations is therefore bounded.   QED.

Note that derivability is decidable in the strong sense of termination of proof search. The search algorithm can be described as follows. Let Γ be the given atoms, and $a = b$ the atom to be derived. First, determine if the conclusion is an instance of *Ref*. If not, apply *Sym* to all atoms of Γ, then *Tr* in all possible ways, and add all new atoms concluded to Γ to get Γ_1. If $a = b$ is not in Γ_1, repeat the above to get Γ_2, and so on. As mentioned in 2.3(c), the number of steps needed in this proof search is polynomial in the number of given atoms.

   The proof of decidability of derivability in the theory of equality extends also to universal formulas, as in Corollary 4.4. An indirect way to show this is as follows. Consider derivations by only the rules of propositional logic, with instances of equality axioms and the given atoms as assumptions. By the subterm property, the instances can be restricted to ones with known terms. Then their number is bounded, and by the decidability of intuitionistic propositional logic, the derivability of universal formulas in the theory of equality is decidable.

(b) Purely syntactic proofs of independence. We are now in a position to prove the independence of rule *Sym* in the theory of equality. If *Sym* were a rule derivable from the other rules, i.e., if its conclusion were derivable from its premiss by the other rules, $b = a$ would be derivable from $a = b$ by *Ref* and *Tr*, which is not the case:

Corollary 2.15. *The atom $b = a$ is not derivable from the assumption $a = b$ in the system of rules Ref + Tr.*

Proof. If there is a derivation, there is one with the subterm property. The only rule with premisses is *Tr*, but any instance of *Tr* with just the terms a, b produces a loop or gives as a conclusion one of $a = a$, $b = b$.   QED.

A standard way of proving the mutual independence of the axioms of a system is to use **models**. A domain of individuals $\mathcal{D}$ is found and an interpretation given to the basic relations of the axiomatic system such that all axioms except one turn out true for the model. For example, let $\mathcal{D} = \mathcal{N}$ (the natural numbers) and consider the relation $n \leqslant m$. It is obviously reflexive and transitive, but not symmetric. Therefore symmetry cannot follow from reflexivity and transitivity.

The idea of proof analysis, as illustrated by the small example of the theory of equality, is to try to see how an axiomatic system works, to get a hold of the structure of derivations, and in particular to give reasons intrinsic to the system for why some derivations are impossible.

2.5 Predicate logic with equality and its word problem

(a) **Replacement rules.** Predicate logic with equality is obtained from standard predicate logic through the addition of a two-place reflexive relation $a = b$ with the property that equals be substitutable everywhere. The latter is formulated as a **replacement axiom**: $A(a)\&a = b \supset A(b)$. (We write for brevity substitutions as $A(a)$ instead of $A(a/x)$, etc.) It is possible to restrict the replacement axiom to atomic predicates and relations. Therefore it is also possible to consider predicate logic with equality as a system of natural deduction extended by two mathematical rules:

Table 2.7 The rules of predicate logic with equality

$$\frac{}{a = a}\ Ref \qquad \frac{P(a) \quad a = b}{P(b)}\ Repl$$

The second rule is schematic: there is one rule for each one-place predicate, and with an n-place relation there is a rule for the replacement of equals for each argument, so altogether there are n rules. These permit us to derive an atom of the form $Q(a, \ldots, c, \ldots, d)$ from the premisses $Q(a, \ldots, b, \ldots, d)$ and $b = c$. It follows at once, by Theorem 2.10, that the normalization of derivations carries over to predicate calculus with equality.

We show first that the equality relation of predicate logic with equality, as defined by the rules of Table 2.7, is an equality relation:

Lemma 2.16. *Rules Sym and Tr are derivable in predicate logic with equality.*

Proof. For *Sym*, set $P(x) \equiv x = a$. The conclusion of *Sym* is derived from its premiss as follows:

$$\frac{\dfrac{}{a = a}\ Ref \quad a = b}{b = a}\ Repl$$

For *Tr*, set $P(x) \equiv x = c$. The conclusion of *Tr* is derived from its premisses as follows:

$$\dfrac{b = c \quad \dfrac{a = b}{b = a}\,{\scriptstyle Sym}}{a = c}\,{\scriptstyle Repl}$$

QED.

We are ready to show that the rule of replacement is admissible for arbitrary formulas:

Lemma 2.17. *Application of replacement to arbitrary formulas reduces to rule Repl.*

Proof. The proof is by induction on the length of the replacement formula. The base case is that of an atomic formula, covered by rule *Repl*. For $\bot$, nothing happens. The other cases are:

1. The formula is $A(a)\&B(a)$. Replacement is reduced to the components $A(a)$ and $B(a)$ as follows:

$$\dfrac{A(a)\&B(a) \quad \dfrac{\dfrac{\overset{1}{A(a)} \quad a = b}{A(b)}\,{\scriptstyle Repl} \quad \dfrac{\overset{1}{B(a)} \quad a = b}{B(b)}\,{\scriptstyle Repl}}{A(b)\&B(b)}\,{\scriptstyle \&I}}{A(b)\&B(b)}\,{\scriptstyle \&E,1}$$

2. With $A(a) \lor B(a)$, the reduction is similar.
3. With $A(a) \supset B(a)$, the reduction is as follows:

$$\dfrac{\dfrac{A(a) \supset B(a) \quad \dfrac{\overset{2}{A(b)} \quad \dfrac{a = b}{b = a}\,{\scriptstyle Sym}}{A(a)}\,{\scriptstyle Repl}}{B(b)} \quad \dfrac{\overset{1}{B(a)} \quad a = b}{B(b)}\,{\scriptstyle Repl}}{A(b) \supset B(b)}\,{\scriptstyle \supset I,2}$$

The quantifiers are treated similarly.   QED.

By this proof, the calculus is complete. Notice that if the standard implication elimination rule were used, the replacement of a with b in the consequent $B(a)$ of $A(a) \supset B(a)$ would have to be done after the logical elimination step, so logical and mathematical parts of derivations could not be maintained apart.

We can give a complete analysis of the role of rule *Ref* in derivations. If *Ref* gives the second premiss of rule *Repl*, a loop is produced. If it gives the first premiss, the replacement predicate is of the form $a = x$ or $x = a$. There are thus altogether three possible steps:

$$\dfrac{P(a) \quad \overline{a = a}^{\,Ref}}{P(a)}\,Repl \qquad \dfrac{\overline{a = a}^{\,Ref} \quad a = b}{a = b}\,Repl \qquad \dfrac{\overline{a = a}^{\,Ref} \quad a = b}{b = a}\,Repl$$

Two cases give a loop, the third the rule of symmetry.

(b) The word problem. We solve the word problem for predicate logic with equality through proof analysis. By the separation of logical rules and instances of rules *Ref* and *Repl*, it is sufficient to consider derivations by the latter two.

Theorem 2.18. *The derivability of an atom from a finite number of given atoms by the rules of predicate logic with equality is decidable.*

Proof. Let Q be derivable from $P_1, \ldots, P_n$ in predicate logic with equality and assume there to be new terms in the derivation. Consider an uppermost such term b, i.e., a new term such that there are no new terms in the subderivation down to the point at which b gets removed. This step is of the form

$$\dfrac{P(b) \quad b = c}{P(c)}\,Repl$$

The second premiss is not an assumption because b is a new term; neither do we have $c \equiv b$ because b would appear in the conclusion. The only possibility that is left is that $b = c$ has been derived by replacement in one of the predicates $x = c$ or $b = x$, but this is impossible: the step would be one of

$$\dfrac{a = c \quad a = b}{b = c}\,Repl \qquad \dfrac{b = a \quad a = c}{b = c}\,Repl$$

The new term is found in an equality in a premiss, which would repeat itself to infinity.　　QED.

Given a finite collection Γ of atoms, the application of *Repl* in minimal derivations produces as conclusions a bounded set of atoms in polynomial time, the **congruence closure** of Γ.

Notes to Chapter 2

Several logicians have suggested generalizations of Gentzen's system of natural deduction. Schroeder-Heister (1984) contains the general version of rule

$\&E$, but an attempt at formalizing a most general rule of implication elimination led to a rule that goes beyond first-order logic. The system of general elimination rules was presented by Tennant (1992) and by Lopez-Escobar (1999), but in both cases with no substantial analysis of the structure of derivations. In particular, the powerful notion of normal derivability, as in Definition 2.4 above, was not used. It turned out that the perhaps first one to have written down the system of rules was Dyckhoff (1988), in whose opinion, at the time, the system was not useful (personal communication). Gentzen's doctoral thesis (1933–34) mentions in passing a proof of normalization for intuitionistic natural deduction. A recently discovered early handwritten version of Gentzen's thesis contains such a proof, written carefully to be ready for publication. An English translation of the chapter on normalization is given in Gentzen (2008). The proof proceeds by first eliminating permutation convertibilities, i.e., premisses of elimination rules derived by $\vee E$ or $\exists E$. Then detour convertibilities are eliminated. See also von Plato (2008, 2009) for the background of the thesis manuscript.

The conversion of universal axioms into rules of proof that extend natural deduction was briefly discussed in Negri and von Plato (1998). As mentioned, similar rules were applied by Prawitz already in 1971, in the proof theory of arithmetic. Even Gentzen's notes from around 1933 have such rules (see von Plato 2009). In Van Dalen and Statman (1978), the axioms of an equality relation are turned into rules added to natural deduction and a proof-theoretical analysis of derivations made.

3 | Axiomatic systems

We shall discuss the organization of an axiomatic system first through an example, namely Hilbert's famous axiomatization of elementary geometry. Hilbert tried to organize the axioms into groups that stem from the division of the basic concepts of geometry such as incidence, order, etc. Next, detailed axiomatizations of plane projective geometry and lattice theory are presented, based on the use of geometric constructions and lattice operations, respectively. An alternative organization of an axiomatic system uses existential axioms in place of such constructions and operations. It is discussed, again through the examples of projective geometry and lattice theory, in Section 3.2.

3.1 Organization of an axiomatization

(a) **Background to axiomatization.** To define an axiomatic system, a language and a system of proof is needed. The language will direct somewhat the construction of an axiomatic system that is added onto the rules of proof: there will be, typically, a domain of individuals, i.e., the objects the axioms talk about, and some basic relations between these objects.

When an axiomatic system is developed in every detail, it becomes a **formal system**. Expressions in the language of the systems are defined inductively, and so are formal proofs. The latter form a sequence that can be produced algorithmically, one after the other.

The idea of a formal axiomatic system is recent, only a hundred years old. Axiomatic systems appeared for the first time in Greek geometry, as known from Euclid's famous book. What he called axioms could be several things: proper axioms, definitions, descriptions of things, and construction postulates. Axioms state things about the objects of geometry and their truth should be immediately evident. The truth of the theorems is to be reduced to the truth of the axioms through proof.

The axiomatization of elementary geometry was a fashionable topic in the late nineteenth century. David Hilbert's book *Grundlagen der Geometrie*

(Foundations of Geometry, 1st edition 1899) became the best known trea-
tise. The first of Hilbert's geometric axioms is in two parts: the first part
states that there is for any two points of the geometric plane a line such that
both points are incident with the line; the second part states that there is at
most one such line.

Table 3.1 Hilbert's first axiom of geometry

I1	For two points A, B there exists always a line a such that both of the points A, B are incident with it.
I2	For two points A, B there exists *not more* that one line with which both points A, B are incident.

Hilbert adds that one always intends expressions such as 'two points' as
'two distinct points'. Mysteriously, Hilbert never refers to his axiom I1 in
the proofs of his book. Instead of mentioning it, he just writes 'AB' for a
line in proofs, whenever two points A and B are given. By axiom I1, there
exists such a line, but the notation is explained nowhere. It turns out that
the formulation of the first axiom as given above was a later one. Hilbert
changed it from the original that reads:

Table 3.2 The original version of Hilbert's first axiom

I1	Two points A, B distinct from each other determine always a line a; we shall set $AB = a$ or $BA = a$.
I2	Any two distinct points of a line determine this line a; that is, if $AB = a$ and $AC = a$, and $B \neq C$, then also $BC = a$.

Hilbert's axiom I1 was originally a construction postulate. Its application in
proofs was clear enough when the object AB appeared. Hilbert didn't care
to change the proofs in the later editions of his book, so the connecting line
construction AB has to be guessed by the reader who knows only a purely
existential formulation of the connecting line axiom. Even so, Hilbert's book
has been hailed as the first one to have really formalized mathematics, to
have dealt with mathematics as a pure game with symbols devoid of intuitive
content.

From Hilbert's axioms and proofs one can gather that he considered
plane geometry with two sorts of objects, points and lines. Notions such as
equal points, equal lines, distinct points, incidence of a point on a line, and
point outside a line are used, but the conceptual order is not given. There

is no explicit system of proof; the axioms are instead applied intuitively. As an extreme case, a proof can be absent, as happens in Hilbert's fifth theorem in the first edition of his book. In the theorem, a connecting line is constructed through two points on distinct sides of a line. In applying the connecting line construction, Hilbert tacitly assumes that the two points are distinct, because it is a condition for the application of the construction. No definition of 'distinct sides' is given. Let us try to figure out what would be needed for a proof of Hilbert's tacit assumption.

Consider a line l, and a point a on one side of the line. For convenience, add an arrowhead to the line so that it has a direction with a appearing on the 'left side' of the line, symbolically, $L(a)$. Let similarly b be a point such that $R(b)$. The task is to show that $a \neq b$. This can be considered a negation of equality, so $\neg a = b$ has to be proved. The following properties of L and R will give a proof:

Table 3.3 Axioms for 'left and right side'

I $\quad \neg(L(a) \,\&\, R(a))$,
II $\quad L(a) \,\&\, a = b \supset L(b), \qquad R(a) \,\&\, a = b \supset R(b)$.

Let us assume $L(a) \,\&\, R(b) \,\&\, a = b$. From $L(a)$ and $a = b$ we get $L(b)$ by II, so $L(b) \,\&\, R(b)$, which contradicts I. Therefore the assumption $a = b$ is wrong and we have proved $\neg a = b$.

The above small example shows what a formal system of proof can do: it can reveal gaps in an axiomatization, such as the lack of explicit principles of substitution of equals in Hilbert's geometry or principles of orientation of the plane. (Such explicit axioms could be found in the work of some of Hilbert's Italian predecessors, for example.)

Our basic language of logic, the predicate calculus, has a sort of ontology of its own: we have a collection of individuals, or of several types of individuals, that make up a domain $\mathcal{D}$ over which the quantifiers range. The predicate calculus can express properties and relations of the individuals of the domain. Equality of two objects is often a crucial relation. An equality $a = b$ means that a and b are expessions for the same object. Say, we write $7 + 5 = 12$ and the meaning is that these different espressions have the same value, i.e., denote the same natural number.

Very many axioms are universal in form. They can be written either as universally quantified formulas, or as formulas with free parameters, as in the following axioms for equality:

Table 3.4 Free-variable and universal axioms

I	$a = a,$	$\forall x.\, x = x,$
II	$a = b \supset b = a,$	$\forall x \forall y.\, x = y \supset y = x,$
III	$a = b \,\&\, b = c \supset a = c,$	$\forall x \forall y \forall y.\, x = y \,\&\, y = z \supset x = z.$

It is convenient to use the notation for free parameters $a, b, c, \ldots$

The notion of distinctness was used by L. Brouwer in the 1920s. His idea seems to have been that the equality of two real numbers needs to be replaced by a notion of 'apartness', with the meaning that $a \neq b$ indicates that a and b 'have a positive distance'. An equality relation can be defined as the negation of apartness, as in $a = b \equiv_{df} \neg a \neq b$. Reflexivity of equality gives the obvious axiom of irreflexivity of the apartness relation, $\neg a \neq a$. Brouwer's second axiom was given an explicit formulation by his student Arend Heyting, and we have:

Table 3.5 Brouwer's axioms for an apartness relation

I	$\neg a \neq a,$
II	$a \neq b \supset a \neq c \vee b \neq c.$

The intuition for the second axiom is very clear: if a and b are apart, if we can't decide $a = c$ because they are 'indefinitely close', then we see that $b \neq c$, and similarly if $b = c$ can't be decided.

By axiom II, we conclude $\neg\, (a \neq c \vee b \neq c) \supset \neg\, a \neq b$, so distributing $\neg$ in the disjunction and using the definition of equality, $a = c \,\&\, b = c \supset a = b$. This is known as 'Euclidean transitivity' from Euclid, who had the axiom: 'any two things equal to a third are equal among themselves'.

Symmetry of apartness is a theorem: substitute a for c in axiom II and the assumption $a \neq b$ gives $a \neq a \vee b \neq a$. The former is excluded by axiom I, so $b \neq a$ remains.

It is important that equals be substitutable in $a \neq b$: if $a \neq b$ and $b = c$, then $a \neq c$. This follows easily because equality was defined in terms of apartness. A similar thing happens if we have a reflexive and transitive partial order relation $a \leqslant b$ and define equality by

$$a = b \equiv_{df} a \leqslant b \,\&\, b \leqslant a$$

(b) Projective geometry. We now give a rigorous axiomatization of plane projective geometry. The structure of the axiomatization is presented in five parts.

1. The domain and its basic relations. We have two sorts of objects in our domain, points denoted a, b, c, ... and lines denoted l, m, n, ... Secondly, we have the basic relations $a = b$, $l = m$, and $a \in l$ (point incident with a line). The last could be written in any way, say $Inc(a, l)$, as no standard notation for incidence has established itself.

After the domain and basic relations are fixed, we consider

I General properties of the basic relations

> *Reflexivity:* $a = a,$ $l = l,$
> *Symmetry:* $a = b \supset b = a$ $l = m \supset m = l,$
> *Transitivity:* $a = b \,\&\, b = c \supset a = c,$ $l = m \,\&\, m = n \supset l = n.$

2. Constructions and their properties. Next we have a choice of geometric constructions and their properties to consider. We shall introduce the **connecting line** of two points a and b, denoted $ln(a, b)$, and the **intersection point** of two lines l and m, denoted $pt(l, m)$. These are formally defined as functions over pairs. Let the domain consist of points denoted Pt and lines denoted Ln. We then have the functions

$$ln : Pt \times Pt \to Ln, \quad pt : Ln \times Ln \to Pt.$$

We can now express the incidence properties of constructed objects as a next group of axioms:

II Properties of constructed objects

> $a = b \vee a \in ln(a, b), \quad a = b \vee b \in ln(a, b),$
> $l = m \vee pt(l, m) \in l, \quad l = m \vee pt(l, m) \in m.$

The axioms state that the line $ln(a, b)$ is constructed exactly through the points a, b, and similarly for the construction $pt(l, m)$. In the axioms, $a = b$ and $l = m$ express **degenerate cases** of the constructions.

3. Uniqueness of constructed objects. We want to have the property that any two points on a line $ln(a, b)$ determine it, i.e., that

> $c \in ln(a, b) \,\&\, d \in ln(a, b) \supset c = d \vee ln(a, b) = ln(c, d),$
> $pt(l, m) \in n \,\&\, pt(l, m) \in k \supset n = k \vee pt(l, m) = pt(n, k).$

A simple formulation is

> $a \in l \,\&\, b \in l \supset a = b \vee ln(a, b) = l,$
> $a \in l \,\&\, a \in m \supset l = m \vee pt(l, m) = a.$

Thoralf Skolem found in 1920 a single axiom from which the uniqueness of both constructions follows, dubbed 'Skolem's beautiful axiom' by Per Martin-Löf:

III Uniqueness of constructions

$$a \in l \,\&\, a \in m \,\&\, b \in l \,\&\, b \in m \supset a = b \vee l = m.$$

The previous formulations follow as special cases of Skolem's axiom.

4. Substitution of equals. We need to guarantee that equals can be substituted in the basic relations. Transitivity of equality is, from this point, just the axiom by which equals are substituted by equals in the equality relations. For the incidence relation, we have the following axioms:

IV Substitution axioms for incidence

$$a \in l \,\&\, a = b \supset b \in l,$$
$$a \in l \,\&\, l = m \supset a \in m.$$

Axiom groups I–IV give the **universal theory** of projective geometry.

5. Existence axioms. To the universal axioms is to be added an axiom of non-collinearity by which there exist at least three non-collinear points:

V Axiom of non-collinearity

$$\exists x \exists y \exists z (\neg x = y \,\&\, \neg z \in ln(x, y)).$$

(c) **Lattice theory.** A certain pattern in the organization of an axiomatization seems to emerge from the above example. We now consider lattice theory under the same pattern. We have a domain $\mathcal{D}$ of individuals $a, b, c, \ldots$ and a partial order over $\mathcal{D}$. Equality is defined by $a = b \equiv a \leqslant b \,\&\, b \leqslant a$. Next we have two operations:

$a \wedge b,$ *the meet of a and b,*
$a \vee b,$ *the join of a and b.*

The axioms are grouped as for projective geometry:

I General properties of the basic relation

Reflexivity: $a \leqslant a,$
Transitivity: $a \leqslant b \,\&\, b \leqslant c \supset a \leqslant c.$

II Properties of constructed objects

$$a \wedge b \leqslant a, \quad a \wedge b \leqslant b,$$
$$a \leqslant a \vee b, \quad b \leqslant a \vee b.$$

III Uniqueness of constructed objects

$$c \leqslant a \,\&\, c \leqslant b \supset c \leqslant a \wedge b,$$
$$a \leqslant c \,\&\, b \leqslant c \supset a \vee b \leqslant c.$$

By axiom III, anything that is 'between' $a \wedge b$ and a and also between $a \wedge b$ and b, is equal to $a \wedge b$:

$$a \wedge b \leqslant c \,\&\, c \leqslant a \,\&\, c \leqslant b \supset c = a \wedge b.$$

The substitution principles for equals in the meet and join constructions are:

$$a = b \supset a \wedge b = c \wedge b, \quad b = c \supset a \wedge b = a \wedge c.$$

These are provable and therefore class IV of substitution axioms is empty.

Lattice theory was born in the latter part of the nineteenth century. One origin was in number theory, in which Richard Dedekind noticed that the greatest common divisor and least common multiple of two natural numbers follow certain abstract laws, namely those for a lattice meet and join. Lattice theory was practised by Ernst Schröder in his 'algebra of logic', though with a terminology and notation that is completely different from that of today. Schröder considered the theory quite abstractly, with various readings of the lattice order relation $a \leqslant b$. The most common reading was that a and b were some sort of domains and the order an inclusion relation, so, in substance, sets with a subset relation. Then meet and join became intersection and union, respectively. In another reading, a and b could be taken as propositions and the order expressed logical consequence with meet and join standing for conjunction and disjunction, or they could be taken as 'circumstances' within a relation of cause and effect. Skolem's early work in logic followed Schröder's algebraic tradition. In 1920, he solved what is today called the word problem for freely generated lattices: A finite number of atomic formulas is assumed given, and the question is what atomic relations these given ones determine. In other words, the problem is to determine, for any atomic formula, if it is derivable from the given atomic formulas. Schröder's terminology and notation were unknown to the extent that Skolem's discovery remained unnoticed until 1992.

A peculiarity of Skolem's axiomatization is that it does not use the lattice operations, but additional basic relations. Why he made this axiomatization is not told to the reader of his article, but it works as a fine illustration of a **relational axiomatization.**

3.2 Relational theories and existential axioms

Relational axiomatizations replace operations with relations. For example, relational lattice theory is based on the idea of having two relations $M(a, b, c)$, $J(a, b, c)$, in addition to the basic order relation, read as *c is the meet of a and b* and *c is the join of a and b*, respectively. There are axioms that state the existence of meets and joins:

$$\forall x \forall y \exists z M(x, y, z), \quad \forall x \forall y \exists z J(x, y, z).$$

There are altogether many more axioms than in an axiomatization with operations, but there are no functions:

I General properties of the basic relations

   Reflexivity: $a \leqslant a$,
   Transitivity: $a \leqslant b \,\&\, b \leqslant c \supset a \leqslant c$.

II Properties of meet and join

   $M(a, b, c) \supset c \leqslant a$, $M(a, b, c) \supset c \leqslant b$,
   $J(a, b, c) \supset a \leqslant c$, $J(a, b, c) \supset b \leqslant c$.

III Uniqueness of meet and join

   $M(a, b, c) \,\&\, d \leqslant a \,\&\, d \leqslant b \supset d \leqslant c$,

   $J(a, b, c) \,\&\, a \leqslant d \,\&\, b \leqslant d \supset c \leqslant d$.

Substitution of equals in the meet and join relations needs to be postulated, with $a = b \equiv a \leqslant b \,\&\, b \leqslant a$. The relations have three arguments, so to cut down the number of axioms, we do as follows:

IV Substitution axioms

   $M(a, b, c) \,\&\, a = d \,\&\, b = e \,\&\, c = f \supset M(d, e, f)$,

   $J(a, b, c) \,\&\, a = d \,\&\, b = e \,\&\, c = f \supset J(d, e, f)$.

Substitution in one argument, say d for a in $M(a, b, c)$, is obtained by the instance

$$M(a, b, c) \mathbin{\&} a = d \mathbin{\&} b = b \mathbin{\&} c = c \supset M(d, b, c).$$

Finally, we have

V Existence of meets and joins

$$\forall x \forall y \exists z M(x, y, z), \quad \forall x \forall y \exists z J(x, y, z).$$

The existential axioms are used as follows:

$$\frac{\dfrac{\forall x \forall y \exists z M(x, y, z)}{\exists z M(a, b, z)} \text{ }_{\forall E, \forall E}}{C} \qquad \frac{\forall x \forall y \exists z M(x, y, z) \quad \begin{array}{c}[M(a, b, v)]\\ \vdots \\ C\end{array}}{C} \text{ }_{\exists E}$$

Here v is an eigenvariable of rule $\exists E$. One would normally use existential axioms by simply considering an instance $M(a, b, v)$ with v arbitrary. This was done by Skolem in 1920, well before Gentzen gave the natural quantifier rules.

Lattice theory with operations has eight axioms, relational lattice theory instead twelve. We show that the former is 'at least as good' as the latter. Define the meet and join relations by

$$M(a, b, c) \equiv a \wedge b = c, \quad J(a, b, c) \equiv a \vee b = c.$$

Axiom V (existence of meets) is derived by

$$\frac{\dfrac{\dfrac{\dfrac{a \wedge b \leqslant a \wedge b \quad a \wedge b \leqslant a \wedge b}{a \wedge b = a \wedge b} \text{ }_{\&I}}{\exists z\, a \wedge b = z} \text{ }_{\exists I}}{\forall x \forall y \exists z\, x \wedge y = z} \text{ }_{\forall I, \forall I}}{}$$

The topformulas are axioms, so, by definition, we have proved the existential axiom $\forall x \forall y \exists z M(x, y, z)$ from the axioms of lattice theory with the meet and join operations. The rest of the relational axioms are derived similarly from the definition of meet and join.

The language of relational lattice theory is not more expressive than that of lattice theory with operations, because the former can be emulated in the latter. In the other direction, to show the equivalence of the two axiomatizations, we proceed as follows. Let an atomic formula $t \leqslant s$ in lattice theory with operations be given. It is translated into relational lattice theory like this: Let a, b be some ground terms in t, with $a \wedge b$ also in t. Take a fresh term c and write down $M(a, b, c)$. If $a \wedge b$ was a component in some lattice term we now have, say, a term $c \wedge d$ and we add $M(c, d, e)$ into our list, with e a fresh term. Proceeding in this way, we find that t is some lattice term, say $e \wedge f$, and we add $M(e, f, g)$ to our list. The same procedure with the term s gives us a list of relational atoms finishing with, say, $M(e' \wedge f', h)$. One now proves by induction on the build-up of terms that if $t \leqslant s$ is provable in lattice theory with operations, then $g \leqslant h$ is provable from $M(a, b, c), \ldots, M(e, f, g), \ldots, M(e', f', h)$ in relational lattice theory.

Let us next try to give a relational axiomatization of plane projective geometry. The relations of equality and incidence will be sufficient. The numbering makes a comparison with the earlier axiomatization direct.

1. The axioms of equality are as before.
2. In place of $a = b \vee a \in ln(a, b)$ we have existence axioms

   $$\forall x \forall y \exists z(x = y \vee (x \in z \,\&\, y \in z)),$$
   $$\forall x \forall y \exists z(x = y \vee (z \in x \,\&\, z \in y)).$$

   The notation is a bit awkward: we read from the two incidence relations $x \in z, y \in z$ in the first axiom that x and y are points and z a line, and dually for the second axiom. One notational possibility is to bound the quantifiers and to write, say, $(\forall x : Pt)$ for a universal quantifier bound to points. Another way is to take more groups of letters into use.

3–4. The axioms of uniqueness and substitution go as before for projective geometry with constructions.

5. The existence of non-collinear points is written

   $$\exists x \exists y \exists z(\neg x = y \vee (\neg x \in z \,\&\, \neg y \in z)).$$

Things go through smoothly in the change to a relational axiomatization. Part of the reason is that we were able to formulate in Section 3.1 the uniqueness and substitution axioms without the geometric constructions.

In a relational formulation, projective geometry has two types of existential axioms, 2 and 5 above.

As with lattice theory, it can be shown that existential axioms are provable from the corresponding axioms for constructions, and that proofs that use constructions can be substituted by proofs that use the existential formulation.

It turns out that theories with operations are clearly easier to treat proof-theoretically than relational theories. Moreover, the two types of existential axioms in projective geometry lead to problems: as is shown in Chapter 8, if one type of axiom is converted to rules, say the existence of connecting lines and intersection points, the other type remains unconvertible, and the other way around.

Notes to Chapter 3

The materials of this chapter come in part from the article von Plato (1996). The first edition of Hilbert's *Grundlagen der Geometrie* can be found, together with Hilbert's lectures that preceded and followed it, in Hallett and Majer (2004). The discovery of the change from constructions to existential axioms was presented first in von Plato (1997). Skolem's work on lattice theory and projective geometry was in practice completely forgotten until it was realized by Stanley Burris, in 1992, that the work on lattice theory contains a polynomial-time solution of the word problem for freely generated lattices. Skolem's relational lattice theory is the topic of Section 5.3. Projective geometry is treated in detail in Chapter 10.

4 | Order and lattice theory

We present in this chapter, first, the theory of partial order. One formulation is based on a weak partial order $a \leqslant b$ and another one on a strict partial order $a < b$. The latter theory is problematic because of the absence of any easy definition of equality. Next, we present lattice theory and give a short, self-contained proof of the subterm property. By this property, we get a solution of the word problem for finitely generated lattices. It also follows that lattice theory is conservative over partial order for the problem of derivability of an atom from given atoms.

In Section 4.3, the most basic structure of algebra, namely a set with an equality and a binary operation, is treated. The proof of the subterm property for such **groupoids** is complicated by the existence of a unit of the operation. The treatment can be generalized to operations with any finite number of terms.

It is possible to modify the rules of lattice theory so that they contain eigenvariables. The number of rules drops down to four instead of six (plus the two of partial order). Moreover, the subterm property has an almost immediate proof. We consider also a formulation of strict order with eigenvariable rules, which permits the introduction (in a literal sense) of a relation of equality. A normal form for derivations and some of its consequences such as the conservativity of strict order with equality over the strict partial order fragment and the subterm property are shown.

4.1 Order relations

(a) **Partial order.** We assume given a domain $\mathcal{D}$ of individuals $a, b, c \ldots$ and a two-place relation $a \leqslant b$ in $\mathcal{D}$ with the following standard axioms:

Table 4.1 The axioms of partial order

PO1   *Reflexivity:* $a \leqslant a$,
PO2   *Transitivity:* $a \leqslant b \,\&\, b \leqslant c \supset a \leqslant c$.

Equality is a defined notion:

Definition 4.1. $a = b \equiv a \leqslant b \& b \leqslant a$.

It is straightforward to convert the two axioms of partial order into rules: these are just like the rules *Ref* and *Tr* of an equality relation. Thus, derivations in the theory of partial order inherit the subterm property of the theory of equality. As with equality, if a premiss of rule *Tr* is a reflexivity, a loop is produced. Therefore the only use of rule *Ref* is to enable us to derive an atom of the form $a \leqslant a$. Otherwise a derivation consists of transitivities that combine a **chain** such as $a_1 \leqslant a_2, a_2 \leqslant a_3, \ldots a_{n-1} \leqslant a_n$ to yield the conclusion $a_1 \leqslant a_n$.

(b) Strict partial order. One may want to use a strict partial order $a < b$ instead of a weak one, with the axioms:

Table 4.2 The axioms of strict partial order

SPO1	Irreflexivity: $\neg a < a$,
SPO2	Transitivity: $a < b \& b < c \supset a < c$.

These axioms convert into the two rules:

Table 4.3 The rules of strict partial order

$$\frac{a < a}{\perp} \, Irr \qquad \frac{a < b \quad b < c}{a < c} \, Tr$$

By Lemma 2.9, the case in which a premiss of *Tr* has been concluded by rule $\perp E$ need not be considered.

Given a set of atoms that form a chain and none of which is of the form $a < a$, rule *Irr* can be applied if the chain contains an inconsistency, exemplified by the mathematical part of the following derivation of the **antisymmetry** of $a < b$:

Table 4.4 An example derivation in strict partial order

$$\cfrac{a < b \& b < a \quad \cfrac{\cfrac{\overset{1}{a < b} \quad \overset{1}{b < a}}{a < c} \, Tr}{\cfrac{a < a}{\perp} \, Irr}}{\cfrac{\perp}{\neg(a < b \& b < a)} \, \supset I, 2} \&E, 1$$

It is not possible to define equality in the theory of strict partial order in the way of weak partial order. If a and b are two elements not compared in either direction in the strict order, the negation of $a < b$ does not mean that $b \leqslant a$. A standard way would be to add equality as a primitive, together with axioms that permit the substitution of equals in the strict order relation:

Table 4.5 The axioms of strict partial order with equality

EQ1	*Reflexivity:* $a = a$,
EQ2	*Symmetry:* $a = b \supset b = a$,
EQ3	*Transitivity:* $a = b \& b = c \supset a = c$,
SUB1	*Substitution:* $a < b \& a = c \supset c < b$,
SUB2	*Substitution:* $a < b \& b = c \supset a < c$.

In derivations by the rules that correspond to these axioms, equality atoms and order atoms mix in rather intricate ways which makes proof analysis hard. We shall see in Section 4.4 that the use of eigenvariable rules results in a neat separation of parts of derivation with rules for strict partial order and rules that permit the introduction of an equality.

4.2 Lattice theory

(a) The subterm property. We consider a system of natural deduction rules for lattice theory with the meet and join operations, as axiomatized in Section 3.1(c):

Table 4.6 The rules of lattice theory in natural deduction style

$$\frac{}{a \leqslant a} \, Ref \qquad \frac{a \leqslant b \quad b \leqslant c}{a \leqslant c} \, Tr$$

$$\frac{}{a \wedge b \leqslant a} \, L\wedge_1 \qquad \frac{}{a \wedge b \leqslant b} \, L\wedge_2 \qquad \frac{c \leqslant a \quad c \leqslant b}{c \leqslant a \wedge b} \, R\wedge$$

$$\frac{}{a \leqslant a \vee b} \, R\vee_1 \qquad \frac{}{b \leqslant a \vee b} \, R\vee_2 \qquad \frac{a \leqslant c \quad b \leqslant c}{a \vee b \leqslant c} \, L\vee$$

Let us call this system of rules **NDLT**. Of the proper rules, $R\wedge$ and $L\vee$ maintain the terms of the premisses in the conclusion, but the middle term in rule Tr is instead lost track of. Transitivity cannot be eliminated in derivations in **NDLT**, but it can be reduced to instances in which the middle term is a subterm of an assumption or of the conclusion. Decidability of the derivability of an atom from given atomic assumptions then follows.

Theorem 4.2. Subterm property for NDLT. *If an atom is derivable from atomic assumptions in* **NDLT**, *it has a derivation with no new terms.*

Proof. Consider a topmost instance of Tr that removes a new term b:

$$\frac{\vdots \qquad \vdots}{a \leqslant b \quad b \leqslant c} \, Tr \\ \frac{}{a \leqslant c}$$

(1)

1. First consider the derivation of the left premiss. If $a \leqslant b$ is concluded by Tr, permute up the Tr removing b:

$$\frac{\dfrac{a \leqslant d \quad d \leqslant b}{a \leqslant b} \, Tr \quad b \leqslant c}{a \leqslant c} \, Tr \quad \rightsquigarrow \quad \frac{a \leqslant d \quad \dfrac{d \leqslant b \quad b \leqslant c}{d \leqslant c} \, Tr}{a \leqslant c} \, Tr$$

(2)

Note that, by assumption, d is not a new term.

If $a \leqslant b$ is concluded by $L\vee$, the term a has a form $a \equiv d \vee e$ and Tr is permuted up as follows:

$$\frac{\dfrac{d \leqslant b \quad e \leqslant b}{d \vee e \leqslant b} \, L\vee \quad b \leqslant c}{d \vee e \leqslant c} \, Tr \quad \rightsquigarrow \quad \frac{\dfrac{d \leqslant b \quad b \leqslant c}{d \leqslant c} \, Tr \quad \dfrac{e \leqslant b \quad b \leqslant c}{e \leqslant c} \, Tr}{d \vee e \leqslant c} \, L\vee$$

(3)

The permutation of Tr removing b is repeated until a left premiss $d' \leqslant b$ is not derived by Tr or $L\vee$. It can be derived by one of the following rules:

1.1. *Ref*: Then $d' \equiv b$ and the right premiss of Tr is identical to the conclusion, so b is not a new term.

$L\wedge_1$: Then $d' \equiv b \wedge e$, so b is not a new term.
$L\wedge_2$: Then $d' \equiv e \wedge b$, so b is not a new term.

1.2. $R\vee_1$: we have $b \equiv d' \vee b'$ and the step

$$\frac{\dfrac{}{d' \leqslant d' \vee b'} \, R\vee_1 \quad d' \vee b' \leqslant c}{d' \leqslant c} \, Tr \\ \vdots \\ a \leqslant c$$

(4)

The case of $R\vee_2$ is similar.

1.3. $R\wedge$: We have some terms a' and d, e such that $b \equiv d \wedge e$ and

$$\frac{\dfrac{a' \leqslant d \quad a' \leqslant e}{a' \leqslant d \wedge e} \, R\wedge \quad d \wedge e \leqslant c}{a' \leqslant c} \, Tr \\ \vdots \\ a \leqslant c$$

(5)

2. Consider the right premiss $b \leqslant c$ of (4) and (5). If concluded by rules *Tr* or $R\wedge$, permute as in (2) and (3).

Rules $R\vee_1$, $R\vee_2$ are excluded dually to the excluded rules $L\wedge_1$, $L\wedge_2$ in the left branch of (1).

This leaves two cases for (4) and also for (5):

2.1. In (4), the right premiss after permutation becomes $d' \vee b' \leqslant c'$ for some term c'

$$
\cfrac{\cfrac{}{d' \leqslant d' \vee b'}\ R\vee_1 \qquad \cfrac{\begin{array}{c}\vdots\end{array}\ \ \cfrac{d' \leqslant c' \quad b' \leqslant c'}{d' \vee b' \leqslant c'}\ L\vee}{}}{\begin{array}{c}d' \leqslant c' \\ \vdots \\ a \leqslant c\end{array}}\ Tr
$$

is transformed into

$$
\begin{array}{c}
\vdots \\
d' \leqslant c' \\
\vdots \\
a \leqslant c
\end{array}
$$

with the transitivity step removed.

2.2

$$
\cfrac{\cfrac{a' \leqslant d \quad a' \leqslant e}{a' \leqslant d \wedge e}\ R\wedge \qquad \cfrac{}{d \wedge e \leqslant c'}\ L\wedge_1}{\begin{array}{c}a' \leqslant c' \\ \vdots \\ a \leqslant c\end{array}}\ Tr \tag{6}
$$

Now $c' \equiv d$ so the derivation is transformed into

$$
\begin{array}{c}
\vdots \\
a' \leqslant d \\
\vdots \\
a \leqslant c
\end{array}
$$

with the transitivity step removed. Rule $L\wedge_2$ is treated similarly.   QED.

The derivability of an atom when a finite number of atoms is assumed is known as the 'word problem for finitely presented lattices'. We have:

Corollary 4.3. Word problem for finitely presented lattices. *The derivability in lattice theory of an atom $a \leqslant b$ from the given assumptions $a_1 \leqslant b_1, \ldots, a_m \leqslant b_m$ is decidable.*

Corollary 4.4. Decidability of universal formulas. *The derivability of universal formulas in lattice theory is decidable.*

Proof. Consider a universal formula in prenex form $\forall x \ldots \forall z A$ with A in conjunctive normal form. Each conjunct A_k is of the form $P_1 \& \ldots \& P_m \supset Q_1 \vee \ldots \vee Q_n$, with P_i, Q_j atoms. The lattice axioms have no disjunctions in positive parts and therefore (by Harrop's theorem, see, e.g., Negri and von Plato 2001) A_k is derivable if and only if $P_1 \& \ldots \& P_m \supset Q_j$ is derivable for some j. Apply Theorem 4.2 to each of the Q_j.   QED.

Terms that do not contain lattice operations will be called **simple terms**. If an atom consists of only simple terms and is derivable in lattice theory, it has by the subterm property a derivation the terms of which are all simple terms, and thus no lattice rule need have been used. Therefore we have:

Corollary 4.5. *Lattice theory is conservative over partial order for universal formulas.*

A proof-theoretical treatment of relational lattice theory, with existential axioms in place of the meet and join operation, is presented in the next chapter.

(b) The Whitman conditions. Consider the following modification of the rules for lattice theory. Rules *Ref, Tr, R*∧, *L*∨ from Table 4.5, and the following:

Table 4.7 Modified rules for lattice theory

$$\frac{a \leqslant c}{a \wedge b \leqslant c} \, L{\wedge}_1' \qquad \frac{b \leqslant c}{a \wedge b \leqslant c} \, L{\wedge}_2' \qquad \frac{c \leqslant a}{c \leqslant a \vee b} \, R{\vee}_1' \qquad \frac{c \leqslant b}{c \leqslant a \vee b} \, R{\vee}_2'$$

The rules of Table 4.6 follow by setting $c \equiv a$ and $c \equiv b$, in turn. In the other direction, the rules of Table 4.7 give those of Table 4.6. For $L{\wedge}_1$, assume the premiss $a \leqslant c$ of the rule in Table 4.7. Its conclusion follows from $a \wedge b \leqslant a$ by *Tr*. Therefore the system of Table 4.7 is equivalent to that of Table 4.6.

   We show that rule *Tr* can be permuted up in the modified system so that at least one of its premisses is an assumption. An immediate proof of the subterm property follows.

Lemma 4.6. *Rule* Tr *can be permuted up in the modified system of lattice theory so that at least one of its premisses is an assumption.*   •

Proof. It is sufficient to consider a derivation with one transitivity as a last rule. If its left premiss is derived by a left rule, Tr is permuted up, and similarly if the right premiss is derived by a right rule. The remaining cases are the following. The left premiss is derived by a right rule and the right by a left rule. We then have, say,

$$\frac{\dfrac{c \leqslant a \quad c \leqslant b}{c \leqslant a \wedge b}\ R\wedge \quad \dfrac{a \leqslant d}{a \wedge b \leqslant d}\ L\wedge_1'}{c \leqslant d}\ Tr$$

This part of the derivation is transformed into

$$\frac{c \leqslant a \quad a \leqslant d}{c \leqslant d}\ Tr$$

The rest of the cases are analogous.   QED.

Theorem 4.7. Subterm property. *If an atom is derivable from atomic assumptions in the modified system of lattice theory, it has a derivation with no new terms.*

Proof. Tr is the only rule that can remove a new term, but by Lemma 4.6, derivations can be so transformed that all middle terms of Tr occur in assumptions.   QED.

It follows especially from Lemma 4.6 that if an atom is derivable with no assumptions, it has a transitivity-free derivation. We then get for free the following result by Phil Whitman (1941) that was among the most important early results of lattice theory:

Corollary 4.8. The Whitman conditions. *If an atom of the form* $a \wedge b \leqslant c \vee d$ *is derivable in lattice theory, one of*

$$a \leqslant c \vee b,\ b \leqslant c \vee d,\ a \wedge b \leqslant c,\ a \wedge b \leqslant d$$

is derivable.

Proof. An inspection of the rules shows that if an atom of the form $a \wedge b \leqslant c \vee d$ is derivable, the last rule has to be one of the four modified rules of Table 4.7.   QED.

4.3 The word problem for groupoids

A **groupoid** is the most basic structure of algebra, in the sense of being the simplest considered: a set with an equality relation and a binary operation with a unit. A solution to the word problem for groupoids gives us a new way of treating the rule of symmetry wherein lies the interest of the problem.

(a) The axioms and rules for a groupoid. We have an unbounded number of terms $a, b, c, \ldots, a_1, b_1, c_1, \ldots$ and a reflexive, symmetric, and transitive equality relation $a = b$. Next we have a **product** ab of two terms, and a **unit** 1. When products are iterated, parentheses are used as in $(ab)c$. The parenthesis notation for products guarantees that if $ab \equiv cd$, then $a \equiv c$ and $b \equiv d$. (Remember that the notation $a \equiv b$ stands for syntactic equality, i.e., that a and b are identical expressions.)

Table 4.8 The axioms for a groupoid

G1	*Equality:*	$a = a,$	$a = b \supset b = a,$	$a = b \,\&\, b = c \supset a = c,$
G2	*Unit:*	$a1 = a,$	$1a = a,$	
G3	*Substitution:*	$a = b \,\&\, c = d \supset ac = bd.$		

These axioms are converted into a rule system:

Table 4.9 Groupoid axioms as a rule system

$$\frac{}{a = a}\,Ref \qquad \frac{a = b}{b = a}\,Sym \qquad \frac{a = b \quad b = c}{a = c}\,Tr$$

$$\frac{}{a1 = a}\,Rl \qquad \frac{}{1a = a}\,Ll$$

$$\frac{a = b \quad c = d}{ac = bd}\,Sub$$

Here R and L indicate products with a unit at right and left, respectively.

(b) The subterm property. We shall now prove the subterm property for derivations of equalities from given equalities.

Definition 4.9. *Let the equalities $a_1 = b_1, \ldots, a_n = b_n$ and $a = b$ be given. Then:*

1. *A term t is **given** if it is a term or subterm in $a_1, b_1, \ldots, a_n, b_n, a, b$.*
2. *A term t is **known** if it is given or of the form $s1$ or $1s$ with s a given term.*

We show that any derivation can be so transformed, through the permutation of the order of application of the rules, that possible new terms get eliminated. By clause 2 in Definition 4.9, we consider known terms modulo product with 1 at left or right, but note that only one multiplication by 1 is permitted. An upper bound is obtained on the number of terms that need to be considered in a word problem, and likewise an upper bound on the number of distinct equalities.

Lemma 4.10. *Rule* Sym *can be permuted up in a derivation so that its premisses are assumptions or conclusions of zero-premiss rules.*

Proof. Consider an uppermost *Sym* in the derivation. Two consecutive instances cancel each other, so there remain two cases. A premiss of *Sym* has been derived by transitivity or by substitution.

1. Transitivity: A part of the derivation and its transformation are

$$\cfrac{\cfrac{a=b \quad b=c}{a=c}\,Tr}{c=a}\,Sym \qquad \cfrac{\cfrac{b=c}{c=b}\,Sym \quad \cfrac{a=b}{b=a}\,Sym}{c=a}\,Tr$$

2. Substitution: A part of the derivation and its transformation are

$$\cfrac{\cfrac{a=b \quad c=d}{ac=bd}\,Sub}{bd=ac}\,Sym \qquad \cfrac{\cfrac{a=b}{b=a}\,Sym \quad \cfrac{c=d}{d=c}\,Sym}{bd=ac}\,Sub$$

A repetition of the above proof transformations brings rule *Sym* up to formulas that start a derivation.   QED.

We shall now add reversed versions of the zero-premiss rules to our system of rules:

Table 4.10 Reversed zero-premiss rules

$$\cfrac{}{a=a1}\,R1rev \qquad \cfrac{}{a=1a}\,L1rev$$

When to the given equalities $a_1 = b_1, \ldots, a_n = b_n$ the reversed equalities $b_1 = a_1, \ldots, b_n = a_n$ are also added, we can leave rule *Sym* out of our system of rules. We then have just two proper rules, i.e., ones with premisses, namely transitivity and substitution in a product. If there is a new term in a derivation, it must be removed before the conclusion is reached, and the only rule that does that is *Tr*.

Theorem 4.11. *A derivation with an instance of* Tr *that removes a new term can be so transformed that no new terms appear.*

Proof. Consider an uppermost new term in the derivation. We permute up the transitivity that removes it so that at least one of its premisses is a topformula, i.e., an assumption or a conclusion of a zero-premiss rule. There are two cases:

1. One premiss is derived by transitivity and the new term is c: a part of the derivation and its transformation are

$$\cfrac{\cfrac{a = b \quad b = c}{a = c}\,Tr \quad c = d}{a = d}\,Tr \qquad \cfrac{a = b \quad \cfrac{b = c \quad c = d}{b = d}\,Tr}{a = d}\,Tr$$

The transformation is similar if the other premiss has been derived by *Tr*.

2. Both premisses are derived by substitution and the new term is bd: a part of the derivation and its transformation are

$$\cfrac{\cfrac{a = b \quad c = d}{ac = bd}\,Sub \quad \cfrac{b = e \quad d = f}{bd = ef}\,Sub}{ac = ef}\,Tr \qquad \cfrac{\cfrac{a = b \quad b = e}{a = e}\,Tr \quad \cfrac{c = d \quad d = f}{c = f}\,Tr}{ac = ef}\,Sub$$

If in 2 b and d are known terms, nothing more needs to be done. Otherwise, repetition of these transformations leads to a situation in which there are no new terms or at least one premiss of the transitivity that removes a new term is a topformula. Then it is an assumption or a conclusion of a zero-premiss rule, but assumptions do not contain new terms. Rule *Ref* produces a loop. In conclusions of the unit rules, at least one side of the equation appears in the conclusion of the transitivity that removes the new term. We have to ensure that the multiplication by 1 does not repeat, as required by clause 2 of Definition 4.9 of known terms.

If one premiss of *Tr* is a unit rule, the other is either a unit rule or *Sub*. In the former case, the term with the unit must be the middle term or else it is a term in the conclusion. The premisses are, say, $b = b1$ and $b1 = b$, but then the conclusion is an instance of *Ref*. Let therefore the other premiss of *Tr* be a conclusion of *Sub*. We then have, say,

$$\cfrac{\cfrac{}{b = b1}\,R1 \quad \cfrac{b = c \quad 1 = d}{b1 = cd}\,Sub}{b = cd}\,Tr$$

Here b is not a given term, so it is either known or a new term. In the latter case, it is removed at some stage by *Tr*. Permutations bring up this *Tr* until we have some left premiss $a = b$, the right premiss $b = cd$ as in the above derivation, and a conclusion $a = cd$. This conclusion is obtained as follows:

$$\dfrac{\dfrac{}{a = a1}\,{\scriptstyle RI} \qquad \dfrac{\dfrac{a = b \quad b = c}{a = c}\,{\scriptstyle Tr} \qquad \dfrac{1 = d}{a1 = cd}\,{\scriptstyle Sub}}{a1 = cd}\,{\scriptstyle Tr}}{a = cd}$$

The term $b1$ has been removed. If b is known but not given, it is of the form $b \equiv a1$ with a a given term. The derivation and its transformation are:

$$\dfrac{\dfrac{}{a1 = (a1)1}\,{\scriptstyle RI} \qquad \dfrac{a1 = c \quad 1 = d}{(a1)1 = cd}\,{\scriptstyle Sub}}{a1 = cd}\,{\scriptstyle Tr} \qquad\qquad \dfrac{\dfrac{\dfrac{}{a = a1}\,{\scriptstyle RI} \quad a1 = c}{a = c}\,{\scriptstyle Tr} \quad 1 = d}{a1 = cd}\,{\scriptstyle Sub}$$

Again, the new term $b1$ has been removed. With an uppermost new term removed, the proof transformation is repeated until no new terms appear.   QED.

The substitution axiom and rule were formulated so that both terms of a product could be substituted simultaneously. This is a crucial property in the above proof. The two axioms $a = b \supset ac = bc$ and $c = d \supset ac = ad$ are together equivalent to $a = b \,\&\, c = d \supset ac = bd$. One learns by trying it that the corresponding two rules would not permit us to make the transformations that are needed for the above theorem.

(c) **Proof search.** The question of the derivability of an equality $a = b$ from given assumptions $a_1 = b_1, \ldots, a_n = b_n$ by the axioms for a groupoid can be effected as follows:

1. Add to the given equations, denoted Γ, the reversed assumptions $b_1 = a_1, \ldots, b_n = a_n$. Let this collection be Γ_1.
2. Add to Γ_1 those conclusions of *Unit* in which at least one side of the equality is a term known from $a_1, b_1, \ldots, a_n, b_n, a, b$. Let this collection be Γ_2.
3. Form the closure of Γ_2 with respect to rules *Tr*, *Sub*, with only known terms in the conclusion of the latter. Let this collection be Γ_3.
4. If $a = b$ is in Γ_3, it is derivable from Γ, otherwise not.

The above proof search procedure can be described in terms of logic programming by turning the axioms for a groupoid into logic programming clauses:

Table 4.11 The axioms for a groupoid as clauses

$\rightarrow a = a,$	$a = b \rightarrow b = a,$	$a = b, b = c \rightarrow a = c,$
$\rightarrow a1 = a,$	$\rightarrow 1a = a,$	$a = b, c = d \rightarrow ac = bd.$

The clause that corresponds to rule *Sym* can be left out by Lemma 4.10.

The word problem can be posed as the question of the derivability of the clause $\Gamma_2 \rightarrow a = b$ through the following rule for the composition of the clauses for a groupoid:

$$\frac{\Gamma \rightarrow a = b \quad a = b, \Delta \rightarrow c = d}{\Gamma, \Delta \rightarrow c = d} \, Comp$$

Note that derivability here has two senses: one for the derivability of an equation, by the rules for a groupoid, another for the derivability of a clause through composition. The subterm property gives that in a derivation of $\Gamma \rightarrow a = b$, the only clause instances that need be used are those with known terms. Further, because all clauses have just one formula in the succedent part, proof search is polynomial by known results of logic programming:

Corollary 4.12. *Proof search for a derivation of an equality from given equalities by the rules for a groupoid has a polynomial upper bound.*

(d) Functions. The above proof of the subterm property for a two-place operation can be generalized to any functions: assume given an equality relation and an n-place function f. Substitution of equals is guaranteed by the axiom

$$a_1 = b_1 \& \ldots \& a_n = b_n \supset f(a_1, \ldots, a_n) = f(b_1, \ldots, b_n).$$

The corresponding rule is

$$\frac{a_1 = b_1 \quad \ldots \quad a_n = b_n}{f(a_1, \ldots, a_n) = f(b_1, \ldots, b_n)} \, Sub$$

Transformation 2 in the proof of Theorem 4.11 generalizes to an arbitrary number of arguments and shows in what way transitivity on function values reduces to n transitivities on the arguments. As in the theorem, it is essential that the substitution is simultaneous.

4.4 Rule systems with eigenvariables

Universally quantified formulas can sometimes be used for defining properties of axiomatic systems otherwise expressible through conditions on constructions. One case is lattice theory in which the meet operation can be characterized axiomatically by the following:

$$\forall x(x \leqslant a \,\&\, x \leqslant b \supset x \leqslant c) \supset \subset a \wedge b \leqslant c.$$

The axiom characterizes the meet of a and b as a greatest element below a and b. The join operation has an analogous characterization.

A weak order can be characterized in an analogous way, through an axiom added to a strict order:

$$\forall x \forall y((x < a \supset x < b) \,\&\, (b < y \supset a < y)) \supset \subset a \leqslant b.$$

Equality can now be defined through the weak order.

We shall convert the above axioms with universal quantifiers into rules that have eigenvariables.

(a) **Lattice theory.** We have the two rules *Ref* and *Tr* of the order relation and the following four rules for the meet and join operations, with x an eigenvariable in rules *LM, RJ*:

Table 4.12 Eigenvariable rules for lattice theory

	$[x \leqslant a \quad x \leqslant b]$	$[a \leqslant x, b \leqslant x]$	
	$\vdots$	$\vdots$	
$\dfrac{c \leqslant a \quad c \leqslant b}{c \leqslant a \wedge b}\,RM$	$\dfrac{x \leqslant d}{a \wedge b \leqslant d}\,LM$	$\dfrac{c \leqslant x}{c \leqslant a \vee b}\,RJ$	$\dfrac{a \leqslant d \quad b \leqslant d}{a \vee b \leqslant d}\,LJ$

These rules are equivalent to the standard lattice rules. Consider rule *LM*: With $d \equiv a$ resp. $d \equiv b$ in LM, the old rules LM_1, LM_2 of Table 4.7 come out as special cases. In the other direction, assume given a derivation of $x \leqslant d$ from the assumptions $x \leqslant a, x \leqslant b$, with x arbitrary. Substitute $a \wedge b$ for x in this derivation, to get a derivation of the conclusion $a \wedge b \leqslant d$ of rule LM from the axioms $a \wedge b \leqslant a, a \wedge b \leqslant b$.

The effect of the eigenvariable rules on the proof of the subterm property is that all problems of permutation of critical transitivities disappear:

Theorem 4.13. *Terms in derivations in lattice theory with eigenvariable rules can be restricted to terms known from assumptions or the conclusion and a bounded number of eigenvariables.*

Proof. Consider the last occurrence of a new term t in a derivation. It is removed by rule Tr and the premisses are $c \leqslant t$ and $t \leqslant d$ for some c, d. As long as at least one premiss is derived by another Tr, the step that removes t is permuted up, and the same if the left premiss is derived by LM or LJ, or the right premiss by RM or RJ. There remain the cases in which the left premiss is derived by RM and the right by LM, and similarly for join. We have $t \equiv a \wedge b$ for some a, b and the part of the derivation

$$\cfrac{\cfrac{c \leqslant a \quad c \leqslant b}{c \leqslant a \wedge b}\,RM \qquad \cfrac{\begin{array}{cc}[x \leqslant a & x \leqslant b]\\ \vdots &\\ x \leqslant d &\end{array}}{a \wedge b \leqslant d}\,LM}{c \leqslant d}\,Tr$$

We take the premisses $c \leqslant a$ and $c \leqslant b$, then substitute c for x in the derivation of $x \leqslant d$ from the assumptions $x \leqslant a$, $x \leqslant b$, to get a derivation of $c \leqslant d$ without the new term $a \wedge b$:

$$\begin{array}{cc} c \leqslant a & c \leqslant b \\ & \vdots \\ & c \leqslant d \end{array}$$

Thus, in the end the instance of Tr that removes the new term disappears, or else at least one of its premisses is an assumption and t not a new term.

Each time a rule with eigenvariables is applied, a lattice operation appears in the conclusion. Therefore the number of lattice operations in the atom to be derived gives an upper bound to the number of eigenvariables in a possible derivation.   QED.

(b) Strict order with equality. We add to the rules of strict partial order of Table 4.3 the following:

Table 4.13 Eigenvariable rules for strict order

$$\cfrac{\begin{array}{cc}[x < a] & [b < y]\\ \vdots & \vdots \\ x < b & a < y\end{array}}{a \leqslant b}\,{\leqslant}I \qquad \cfrac{a \leqslant b \quad c < a}{c < b}\,{\leqslant}E_1 \qquad \cfrac{a \leqslant b \quad b < c}{a < c}\,{\leqslant}E_2$$

In rule ${\leqslant}I$, x and y are distinct eigenvariables, not free in the conclusion nor in any other assumptions than those shown as discharged in the schematic rule.

Theorem 4.14. *The order relation $a \leqslant b$ defined by the rules of Table 4.13 is reflexive and transitive.*

Proof. The following derivations show that *Ref* and *Tr* for weak partial order are derivable rules:

1. Reflexivity:

$$\frac{[x < a] \quad [a < y]}{a \leqslant a} \leqslant I$$

The derivations of the two premisses of rule $\leqslant I$ are degenerate.

2. Transitivity:

$$\frac{b \leqslant c \quad \dfrac{a \leqslant b \quad [x < a]}{x < b} \leqslant E_1}{\dfrac{x < c}{a \leqslant c}} \leqslant E_1 \qquad \frac{a \leqslant b \quad \dfrac{b \leqslant c \quad [c < y]}{b < y} \leqslant E_2}{\dfrac{a < y}{} \leqslant E_2} \leqslant I$$

QED.

Definition 4.15. Equality. $a = b \equiv a \leqslant b \,\&\, b \leqslant a$.

Theorem 4.16. *The relation $a = b$ of Definition 4.15 is an equivalence relation.*

Proof. Reflexivity and transitivity follow by Theorem 4.14 and symmetry by Definition 4.15.   QED.

Rules $\leqslant E_1$ and $\leqslant E_2$ guarantee that equals can be substituted in the strict order relation, so the axioms of Table 4.5, the one with a basic strict order and equivalence relations and substitution axioms for equals in the order relation, are all provable.

Definition 4.17. Normal derivation. *A derivation in strict order with equality is **normal** if no instance of rule $\leqslant I$ is followed by rule $\leqslant E$.*

Lemma 4.18. *Derivations in strict order with equality convert to normal form.*

Proof. Consider a derivation that is not normal. It has at least one pair of rules $\leqslant I, \leqslant E_1$ or $\leqslant I, \leqslant E_2$. With the second, the part of derivation is

$$
\cfrac{\cfrac{\begin{array}{cc}[x < a] & [b < y]\\ \vdots & \vdots \\ x < b & a < y\end{array}}{a \leqslant b} \leqslant I \qquad \begin{array}{c}\vdots\\ b < c\end{array}}{a < c} \leqslant E_2
$$

$$\vdots$$

This part is converted, by the substitution of c for the arbitrary y in the derivation of the second premiss of $\leqslant I$, into

$$
\begin{array}{c}
\vdots \\
b < c \\
\vdots \\
a < c \\
\vdots
\end{array}
$$

The case is similar for the first pair of convertible rules. Repetition of these conversions removes all I-E pairs from derivations.   QED.

We give an example derivation:

$$
\cfrac{\cfrac{\overset{1}{[x < a]} \quad \overset{2}{[a < b]}}{x < b} {\scriptstyle Tr} \qquad \cfrac{\overset{2}{[a < b]} \quad \overset{1}{[b < y]}}{a < y} {\scriptstyle Tr}}{\cfrac{a \leqslant b}{a < b \supset a \leqslant b}\,{\scriptstyle \supset I,2}}{\scriptstyle \leqslant I,1}
$$

Let us call atoms of the form $a < b$ **S-atoms**, and those of the form $a \leqslant b$ **W-atoms**.

Theorem 4.19. Subterm property. *If an S-atom is derivable from given atoms in strict order with equality, it has a derivation in which all terms are known. If a W-atom is similarly derivable, it has a derivation in which all terms are known except for two eigenvariables x, y of the last rule of the derivation.*

Proof. We can assume the derivation to be normal. We show first how to reduce the derivability of W-atoms to the derivability of S-atoms.

The only rule that allows us to derive W-atoms is $\leqslant I$. Thus, if the conclusion is a W-atom $a \leqslant b$ and the assumptions Γ, we have subderivations of the premisses of a last rule $\leqslant I$ as in

$$
\begin{array}{cc}
\Gamma', x < a & \Gamma'', b < y \\
\vdots & \vdots \\
x < b & a < y
\end{array}
$$

Here $\Gamma', \Gamma'' \equiv \Gamma$, and $a \leqslant b$ is derivable from Γ by the rules of strict order with equality if and only if $x < b$ is derivable from $x < a$ and assumptions Γ' contained in Γ and $a < y$ derivable from $b < y$ and assumptions Γ'' contained in Γ.

Next assume the conclusion to be an S-atom $a < b$. The derivation begins with instances of Tr and $\leqslant E$, but the latter have a major (left) premiss that is an assumption. Consider a new term removed by Tr, and permute it up until at least one of its premisses is derived by $\leqslant E$, as in

$$
\cfrac{\cfrac{c \leqslant d \quad \begin{array}{c} \vdots \\ d < e \end{array}}{c < e}\leqslant E_2 \qquad \begin{array}{c} \vdots \\ e < f \end{array}}{c < f}\ Tr
$$
$$
\vdots
$$

Here e is a new term. The rule that removes it is permuted up:

$$
\cfrac{c \leqslant d \quad \cfrac{\begin{array}{c} \vdots \\ d < e \end{array} \quad \begin{array}{c} \vdots \\ e < f \end{array}}{d < f}\ Tr}{c < f}\leqslant E_2
$$
$$
\vdots
$$

In the end, the rule that removes the new term has an assumption as one premiss, which is impossible. Therefore the subterm property for the derivations of S-atoms follows. By the reduction of derivations of W-atoms to those of S-atoms, the second part of the theorem follows.   QED.

A rule system that corresponds to a standard axiomatization of strict order and equality, as in Table 4.5, produces derivations in which there is a mess of order and equality atoms, hard to analyse combinatorially. With the eigenvariable rule, the corresponding parts of derivations with S- and W-atoms are neatly separated. In particular, if no assumptions are of the form

$a \leqslant b$, and neither the conclusion, a normal derivation has no instances of rules $\leqslant I, \leqslant E$. Therefore we have the following corollary.

Corollary 4.20. *Strict order with equality is conservative over strict partial order for the derivation of atoms from atomic assumptions.*

Notes to Chapter 4

The solution of the word problem for freely generated lattices in Section 4.2 comes from Negri and von Plato (2002). The Whitman conditions are derived through proof analysis in Negri and von Plato (2004). (An annoying slip of pen in corollary 6.4 of that paper claimed the derivability of one of $a \leqslant b, a \leqslant c, b \leqslant c, b \leqslant d$ whenever $a \wedge b \leqslant c \vee d$ is derivable.)

The rule system for groupoids of Section 4.3 is new. We did not know anything about a possible solution in more algebraic terms but just did the example as an illustration of the proof-theoretical method in algebra; it would be surprising if such a solution did not exist.

Rule systems with eigenvariables for lattice theory and the solution of the word problem were given in the historical paper von Plato (2007). The rule system for strict order with equality with its introduction and elimination rules for weak partial order is new.

Our proof-theoretical solution to the word problem for freely generated lattices has been carried further by Andrea Meinander in her master's thesis (Helsinki 2007), to solve the corresponding problem for **ortholattices,** a problem that seems to have resisted solution by more traditional methods of lattice theory (cf. Meinander 2010).

5 | Theories with existence axioms

We show how to extend natural deduction systematically by rules that correspond to existential axioms. Then we treat, as simple examples, theories of equality and order in which conditions of non-triviality or non-degeneracy have been added. Last, we give a detailed proof analysis of derivations in relational lattice theory.

5.1 Existence in natural deduction

So far we have extended natural deduction by rules that correspond to universal axioms, and in the last section of the previous chapter, by rules that correspond to axioms in which a universal formula implies an atomic one. We shall in this chapter consider existential axioms of the form

$$\forall x_1 \ldots \forall x_n \exists y_1 \ldots \exists y_m A.$$

Certain restrictions will be put on formula A, supposed to be quantifier free. An axiom of the above form would be used, in natural deduction, as in the following schematic derivation:

$$\cfrac{\cfrac{\forall x_1 \ldots \forall x_n \exists y_1 \ldots \exists y_m A}{\exists y_1 \ldots \exists y_m A(a_1/x_1, \ldots, a_n/x_n)} \,{\scriptstyle \forall E \ldots \forall E} \quad \cfrac{[\exists y_m A(a_1/x_1, \ldots, a_n/x_n, z_1/y_1, \ldots, z_{m-1}/y_{m-1})] \quad \cfrac{[A(a_1/x_1, \ldots, a_n/x_n, z_1/y_1, \ldots, z_m/y_m)]}{\vdots}}{\cfrac{C}{\vdots \; C}} \,{\scriptstyle \exists E}}{C} \,{\scriptstyle \exists E}$$

This logical derivation from the axiom $\forall x_1 \ldots \forall x_n \exists y_1 \ldots \exists y_m A$ can be written as a rule from which the quantifier elimination steps have been removed:

$$\cfrac{[A(a_1/x_1, \ldots, a_n/x_n, z_1/y_1, \ldots, z_m/y_m)]}{\cfrac{\vdots}{\cfrac{C}{C}}}$$

Here $a_1, \ldots a_n$ are free parameters and $z_1, \ldots z_m$ eigenvariables, and the assumption is closed at the inference. It depends on the form of A whether a logic-free rule can be formulated.

A very general class of axioms convertible to rules is formed by what are called in category theory **geometric implications**. They are defined as follows. First, if A contains no $\supset$ nor $\forall$, it is a **geometric formula**. (Note that we treat negation in natural deduction as a special case of implication.) Secondly, we set the following definition:

Definition 5.1. Geometric implication. *If A and B are geometric formulas, then formulas of the form*

$$\forall x_1 \ldots \dot{\forall} x_n (A \supset B)$$

are **geometric implications**.

Geometric implications can be written in the following equivalent way: they are conjunctions of universal closures of formulas of the form

$$P_1 \& \ldots \& P_m \supset \exists y_{1_1} \ldots \exists y_{1_k} M_1 \vee \ldots \vee \exists y_{n_1} \ldots \exists y_{n_l} M_n$$

Here the P_i are atoms, the M_j conjunctions of atoms, and the variables y do not occur in the P_i.

The use of natural deduction poses a restriction on axioms that can be transformed into rules of inference. The condition was, as in Definition 2.8, that there be no disjunctions in positive parts of formulas. The corresponding restriction in geometric implications is met in the special case of $n = 1$ and $M \equiv Q_1 \& \ldots \& Q_s$ that gives

$$P_1 \& \ldots \& P_m \supset \exists y_1 \ldots \exists y_k (Q_1 \& \ldots \& Q_s).$$

We then arrive at the following natural deduction formulation of the **geometric rule scheme**:

$$
\frac{P_1 \quad \ldots \quad P_m \qquad \overset{\displaystyle [Q_1, \ldots, Q_s]}{\underset{\displaystyle C}{\vdots}}}{C}
$$

Here the assumptions $Q_1, \ldots, Q_s$ are closed at the inference. These assumptions may contain eigenvariables $v_1, \ldots, v_k$ that must not occur free in C nor in any other assumptions.

It is easily seen that a geometric rule has the same deductive strength as the corresponding axiom. The way from the axiom to the rule goes as in the big schematic derivation above. The other way goes by forming the conjunction of the assumptions Q_i, then quantifying existentially to get $\exists y_1 \ldots \exists y_k (Q_1 \& \ldots \& Q_s)$ as the minor premiss of the geometric rule. When its major premisses $P_1, \ldots, P_m$ are concluded from $P_1 \& \ldots \& P_m$, the geometric rule gives the conclusion $\exists y_1 \ldots \exists y_k (Q_1 \& \ldots \& Q_s)$ and implication introduction the axiom.

If a major premiss P_i of a geometric rule has been concluded by a logical rule, it must have been an elimination rule, and a permutation as in the proof of Theorem 2.9 can be applied.

Next to the geometric axioms and rules, we note the possibility of a dual **co-geometric** class of axioms and rules. A formula A is defined to be **co-geometric** if A contains no $\supset$ nor $\exists$.

Definition 5.2. Co-geometric implication. *If A and B are co-geometric formulas, then formulas of the form*

$$\forall x_1 \ldots \forall x_n (A \supset B)$$

are **co-geometric implications.**

Co-geometric implications can be written in the following equivalent way: They are conjunctions of universal closures of formulas of the form (with $\bar{x}_i$ vectors of variables)

$$\forall z_1 \ldots \forall z_n (\forall \bar{x}_1 (Q_{1_1} \vee \ldots \vee Q_{1_k}) \& \ldots \& \forall \bar{x}_n (Q_{n_1} \vee \ldots \vee Q_{n_l}) \supset P_1 \vee \ldots \vee P_m).$$

Variable conditions are dual to those for the geometric implications.

The cases that can be converted to rules in natural deduction style are, for geometric and co-geometric implications, as follows:

Table 5.1 Geometric and co-geometric rules in natural deduction

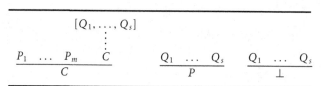

In typical cases of geometric and co-geometric implications, the atoms Q_i contain eigenvariables that correspond to the existential quantifiers. A co-geometric implication can be converted into a rule in natural deduction style only if $m \leq 1$. The two cases $m = 1$ and $m = 0$ give as conclusions an atom P and $\bot$, respectively.

The rules act only on atomic formulas, but variable conditions can prevent the permutation of logical rules to below the mathematical ones. However, it happens in typical cases that rules with variable conditions are conservative over the rest of the rules for the derivation of atoms from given atoms.

5.2 Theories of equality and order again

We shall show by simple examples how existential axioms work as rules of inference, and what type of results can be obtained by their use.

(a) **Non-triviality in equality.** The theory of equality would become trivial in a case in which all elements are equal. Thus we want to add an axiom that dictates the existence of at least two unequal elements. The condition of **non-triviality** can be put as a fourth axiom of equality, next to those of reflexivity, symmetry, and transitivity:

> EQ4 **Non-triviality:** $\exists x \exists y \neg x = y.$

The axiom is different from those converted to rules so far, but it is classically equivalent to a co-geometric implication with $m = 0$ and can be formulated as a special case of the co-geometric rule scheme:

$$\frac{x = y}{\bot} \; Ntriv$$

Here x and y are the eigenvariables of the rule, assumed distinct and not free in any assumptions the premiss of the rule depends on. The idea is that if one can prove $x = y$ for arbitrary x and y, one can conclude anything.

As mentioned, rules with variable conditions can often be put aside in word problems. For non-trivial equality, we have:

Theorem 5.3. *No premiss of* Ntriv *is derivable from atomic assumptions by the rules of equality.*

Proof. Consider a derivation of an atom from given atoms by rules *Ref, Sym,* and *Tr.* No such atom can work as the premiss of rule *Ntriv*: a topformula in the derivation can contain x or y only if the formula comes from an instance of *Ref,* but $x = x$ and $y = y$ produce a loop if rule *Sym* or *Tr* is applied, and rule *Ntriv* is excluded by the requirement that x and y be distinct.   QED.

By the theorem, *Ntriv* gives a conservative extension of the theory of equality as far as the derivability of atoms from given atoms is concerned. The proper use of existential axioms, such as EQ4, requires a conclusion with a logical structure.

(b) Non-degenerate partial order. We can add to strict partial order an axiom that is similar to the one for non-trivial equality:

SPO3 **Non-degeneracy:** $\exists x \exists y\, x < y.$

This axiom is a geometric implication and can be converted into the geometric rule without major premisses:

$$
\begin{array}{c}
[x < y] \\
\vdots \\
\dfrac{C}{C}\,Ndeg
\end{array}
$$

Here x and y are the eigenvariables of the rule, assumed distinct and not free in any assumptions the premiss of the rule depends on except $x < y$ that is closed by the application of the rule.

Theorem 5.4. *If $a < b$ is derivable from $a_1 < b_1, \ldots, a_n < b_n$ in the theory of non-degenerate strict partial order, it is derivable without rule* Ndeg, *and similarly if $\perp$ is derivable from $a_1 < b_1, \ldots, a_n < b_n$.*

Proof. Assume there are instances of *Ndeg* in a derivation. We transform the derivation so that this turns out impossible. If the conclusion of an uppermost instance of *Ndeg* is a premiss of *Tr*, the two rules can be permuted as in

$$
\begin{array}{c}
[x < y] \\
\vdots \\
\dfrac{c < d}{c < d}\,Ndeg \quad d < e \\
\dfrac{\phantom{c < d}}{c < e}\,Tr
\end{array}
\qquad
\begin{array}{c}
[x < y] \\
\vdots \\
\dfrac{c < d \quad d < e}{c < e}\,Tr \\
\dfrac{\phantom{c < e}}{c < e}\,Ndeg
\end{array}
$$

Eigenvariables of other possible instances of rule *Ndeg* in the derivation of $d < e$ do not interfere because all eigenvariable rules have their own distinct variables. Therefore a derivation can be so transformed that all eigenvariable rules are applied last.

Now consider a derivation with applications of *Ndeg* as last rules. The subderivation with just the uppermost *Ndeg* is:

$$\Gamma, [x < y]$$
$$\vdots$$
$$\frac{a < b}{a < b} \, Ndeg$$

Here Γ denotes those given assumptions $a_i < b_i$ that were used in the derivation and possible assumptions with eigenvariables in rules that come after the uppermost instance of *Ndeg* and that were also used in the derivation of $a < b$. By the subterm property for minimal derivations in strict partial order, all terms in the derivation of the premiss of *Ndeg* are known. The only way to use the assumption $x < y$ is in rule *Tr*, but then either x or y would have to appear in an assumption other than $x < y$, which is impossible because of the requirement that each eigenvariable rule close only assumptions with its own eigenvariables and nothing else. Therefore $x < y$ was used 0 times in the derivation, and the instance of *Ndeg* can be dropped as superfluous.

After the uppermost *Ndeg* has been deleted, the rest follow one after the other.

If $\bot$ is derivable from $a_1 < b_1, \ldots, a_n < b_n$ with *Ndeg* as a last rule, there is an instance of *Irref* above *Ndeg*:

$$\Gamma, [x < y]$$
$$\vdots$$
$$\frac{a < a}{\dfrac{\bot}{\bot}}$$

with *Irref* and *Ndeg* labels.

As above, the formula $x < y$ could be used only in *Tr*, but then either x or y would appear in an atom that cannot be closed, by which the instance of *Ndeg* would be superfluous.   QED.

5.3 Relational lattice theory

We study a system of rules for lattice theory that corresponds to Skolem's original work (1920). The rules are given in a single-conclusion formulation in natural deduction style.

(a) The rules of relational lattice theory. The relational axiomatization of lattice theory, presented in Section 3.2, uses existence axioms for meets and joins instead of explicit meet and join operations. We assume an infinity of parameters $a, b, c, \ldots$ and variables $x, y, z, \ldots$ There is a binary partial order relation $a \leqslant b$ and two ternary relations $M(a, b, c)$ and $J(a, b, c)$ ('c is the meet of a and b', and 'c is the join of a and b'). We call atoms of these forms **O-atoms**, **M-atoms**, and **J-atoms**. Equality is partial order in both directions. In the substitution rules below, we abbreviate the two premisses $a \leqslant b$ and $b \leqslant a$ by $a = b$. The first rule has zero premisses. In rules III–IV, the mnemonic letters L and R indicate that the meet and join terms appear as left and right members, respectively, of the order relation in the conclusion. The rules for lattice theory, with the roman numerals signalling correspondence with the axiomatization in Section 3.2, are:

Table 5.2 Rules for relational lattice theory ReLT

I–II *Rules for partial order:*

$$\frac{}{a \leqslant a}\ Ref \qquad\qquad \frac{a \leqslant b \quad b \leqslant c}{a \leqslant c}\ Tr$$

III–IV *Rules for Meet and Join:*

$$\frac{M(a, b, c)}{c \leqslant a}\ LM_1 \qquad \frac{M(a, b, c)}{c \leqslant b}\ LM_2 \qquad \frac{J(a, b, c)}{a \leqslant c}\ RJ_1 \qquad \frac{J(a, b, c)}{b \leqslant c}\ RJ_2$$

$$\frac{M(a, b, c) \quad d \leqslant a \quad d \leqslant b}{d \leqslant c}\ RM \qquad \frac{J(a, b, c) \quad a \leqslant d \quad b \leqslant d}{c \leqslant d}\ LJ$$

V *Substitution of equals in Meet and Join:*

$$\frac{M(a, b, c) \quad a = d \quad b = e \quad c = f}{M(d, e, f)}\ SM \qquad \frac{J(a, b, c) \quad a = d \quad b = e \quad c = f}{J(d, e, f)}\ SJ$$

VI *Existential rules for Meet and Join:*

$$\frac{\begin{array}{c}[M(a, b, x)]\\ \vdots\\ C\end{array}}{C}\ EM \qquad\qquad \frac{\begin{array}{c}[J(a, b, x)]\\ \vdots\\ C\end{array}}{C}\ EJ$$

A derivation can begin with any O-, M-, or J-atoms as assumptions. The existential rules have the variable restriction that the eigenvariable x must not occur free in the conclusion C nor in any open assumption C depends on, except in the M- or J-atoms indicated as closed. We assume that the eigenvariable of a rule appears only in the subderivation down to that rule. It follows that all the eigenvariables of existential rules in a derivation are distinct.

The existential rules are equivalent to the existence axioms for meet and join, namely $\forall x \forall y \exists z M(x, y, z)$ and $\forall x \forall y \exists z J(x, y, z)$. If the latter are assumed, the logical rules of universal and existential quantifier elimination lead to the conclusions of the existential lattice rules, and in the other direction, the existence axioms are derivable by universal and existential quantifier introduction and the existential lattice rules:

$$
\frac{\forall x \forall y \exists z M(x, y, z)}{\exists z M(a, b, z)} \ {\scriptstyle \forall E, \forall E}
\qquad
\frac{
\begin{array}{c}
[M(a, b, v)] \\
\vdots \\
C
\end{array}
}{C} \ {\scriptstyle \exists E}
\qquad
\frac{\dfrac{\dfrac{[M(a, b, v)]}{\exists z M(a, b, z)} \ {\scriptstyle \exists I}}{\exists z M(a, b, z)} \ {\scriptstyle EM}}{\forall x \forall y \exists z M(x, y, z)} \ {\scriptstyle \forall I, \forall I}
$$

From the left derivation, we observe that an existence axiom turns into a corresponding existential rule of inference by the deletion of the existential premiss. From the right derivation, we observe that the existential rule is applied after rule $\exists I$ and that this order cannot be permuted. However, there is an existential quantifier in the conclusion and therefore this lack of permutability does not influence the derivability of atomic formulas.

In Skolem (1920), rules I–V are treated formally, but existence axioms and their variable restrictions are handled somewhat intuitively.

(b) Permutability of rules. The order of application of lattice rules can be permuted by suitable local transformations:

Lemma 5.5. (i) *Instances of rules EM, EJ permute down with respect to all the other rules of* **ReLT**. (ii) *Instances of rules SM, SJ permute down with respect to all the rules except for EM, EJ. If the conclusion is an O-atom, no instance of SM, SJ is needed, and otherwise just one instance of SM, SJ is sufficient.*

Proof. (i) If *EM* or *EJ* enables us to conclude an atom C and C is a premiss of a lattice rule R that lets us conclude D, rule R is applied to the conclusion C of *EM* or *EJ*, and then *EM* or *EJ* is applied to D. By the conditions on eigenvariables, this can always be done. (ii) Consider a substitution on a in $M(a, b, c)$. We can leave out the superfluous premisses $b = b$ and $c = c$ and have the instance

$$
\frac{M(a, b, c) \quad a \leqslant d \quad d \leqslant a}{M(d, b, c)} \ {\scriptstyle SM}
$$

The conclusion $M(d, b, c)$ can be a premiss in $L M_1$, $L M_2$, and RM. In the first case, make the conversion

$$
\frac{\dfrac{M(a, b, c) \quad a \leqslant d \quad d \leqslant a}{M(d, b, c)} \ {\scriptstyle SM}}{c \leqslant d} \ {\scriptstyle L M_1}
\qquad
\frac{\dfrac{M(a, b, c)}{c \leqslant a} \ {\scriptstyle L M_1} \quad a \leqslant d}{c \leqslant d} \ {\scriptstyle Tr}
$$

In the case of LM_2 we convert as follows:

$$\dfrac{\dfrac{M(a, b, c) \quad a \leqslant d \quad d \leqslant a}{M(d, b, c)} \, SM}{c \leqslant b} \, LM_2 \qquad\qquad \dfrac{M(a, b, c)}{c \leqslant b} \, LM_2$$

If $M(d, b, c)$ is a premiss in RM, the derivation is

$$\dfrac{\dfrac{M(a, b, c) \, a \leqslant d \, d \leqslant a}{M(d, b, c)} \, SM \qquad e \leqslant d \quad e \leqslant b}{e \leqslant c} \, RM$$

This is converted into

$$\dfrac{M(a, b, c) \qquad \dfrac{e \leqslant d \quad d \leqslant a}{e \leqslant a} \, Tr \qquad e \leqslant b}{e \leqslant c} \, RM$$

Other cases of substitutions are variants of these three, until when permuting down substitution another substitution is met. We have, again assuming substitutions on the first argument:

$$\dfrac{\dfrac{M(a, b, c) \quad a \leqslant d \quad d \leqslant a}{M(d, b, c)} \, SM \qquad d \leqslant e \quad e \leqslant d}{M(e, b, c)} \, SM$$

This is converted into transitivities and one substitution:

$$\dfrac{M(a, b, c) \qquad \dfrac{a \leqslant d \quad d \leqslant e}{a \leqslant e} \, Tr \qquad \dfrac{e \leqslant d \quad d \leqslant a}{e \leqslant a} \, Tr}{M(e, b, c)} \, SM$$

No variable restrictions are violated by the above proof transformations, so the transformations give a correct derivation of the original conclusion. In the end, if the conclusion is an O-atom, no substitutions are needed, and otherwise there is at most one substitution as a last rule.   QED.

Lemma 5.5 corresponds to lemma 2 in Skolem (1920). Rules SM, SJ, EM, EJ are the only ones that permit concluding M- or J-atoms. If existential rules are permuted down and if the conclusion of the derivation is an O-atom, no substitutions are needed down to the derivation of the premiss of the first existential rule, and therefore no substitutions at all. We show later that derivations that end with M- or J-atoms can be reduced to derivations that enable us to derive O-atoms, so, by Lemma 5.5, we do not need to consider rules SM, SJ.

Definition 5.6. *A derivation tree in* **ReLT** *is* **loop free** *if it has no branches in which the same atom occurs more than once, except as a premiss and conclusion of an existential rule, and atoms of the form $a \leqslant a$ appear only as leaves. A term in a derivation tree that is not a term in an open assumption or the conclusion is a* **new term**.

Lemma 5.7. *In a loop-free derivation of an O-atom with no instances of rules EM, EJ, there are no new terms in the derivation.*

Proof. We may assume by Lemma 5.5 that there are no instances of rules *SM, SJ*. Then *M*- and *J*-atoms are never conclusions, so terms in them remain terms in open assumptions. Rule *Tr* is the only one that can remove a new term, say b:

$$\frac{a \leqslant b \quad b \leqslant c}{a \leqslant c} \, Tr$$

Trace up atoms with the new term. First occurrences of b cannot be in any *M*- or *J*-atoms or other assumptions. Thus, the new term must appear first in instances $b \leqslant b$ of rule *Ref*. Such an instance is not a premiss of *Tr* because the conclusion would be equal to the other premiss and the derivation would have a loop. Therefore $b \leqslant b$ is a premiss of *RM* or *LJ*, say

$$\frac{M(b, e, f) \quad b \leqslant b \quad b \leqslant e}{b \leqslant f} \, RM \qquad \frac{J(b, e, f) \quad b \leqslant b \quad e \leqslant b}{f \leqslant b} \, LJ$$

Now the new term is in an *M*- or *J*-atom, contrary to assumption, and similarly if $b \leqslant b$ is the last premiss of *RM* or *LJ*.   QED.

Lemma 5.8. *In a loop-free derivation of an O-atom with one existential rule EM or EJ as a last step and closed atom $M(a, b, v)$ or $J(a, b, v)$, first occurrences of the eigenvariable v are not in instances of rule Ref.*

Proof. By Lemma 5.5, the derivation of the *O*-atom premiss of rule *EM* does not need rules *SM, SJ*. Assume there is a leaf in the derivation tree that begins with $v \leqslant v$. It is not a premiss in *Tr* or there is a loop. By the variable restriction on rule *EM*, v is not in any open assumption. Therefore $v \leqslant v$ is not a premiss in rule *LJ*. So $v \leqslant v$ is a premiss in *RM*, but then the first or second argument in $M(a, b, v)$ is v and the conclusion of *RM* is the same as the premiss $v \leqslant v$. The proof for *EJ* is dual to above.   QED.

Theorem 5.9. Subterm property. *If an O-atom is derivable from atomic assumptions in* **ReLT**, *it has a derivation with no new terms.*

Proof. We may assume that the derivation is loop free. If there are no instances of *EM* or *EJ*, the result is given in Lemma 5.7. We show that derivations with existential rules can be transformed through suitable permutations into ones with loops. Assume the derivation has instances of *EM* or *EJ*. By Lemma 5.5, these can be permuted last, and each of them enables us to derive the *O*-atom that is the conclusion of the whole derivation. Consider the subderivation down to a first instance of an existential rule, say *EM*, that closes the assumption $M(a, b, v)$. By Lemma 5.5, rules *SM*, *SJ* can be assumed absent, so all *M*- and *J*-atoms in the derivation are assumptions. The eigenvariable v is a new term and by Lemma 5.8, all topmost occurrences of v are in the closed assumptions $M(a, b, v)$. We transform the derivation into another one that has the same terms and show that either it has a loop or else it has the subterm property. The transformation consists in permuting up instances of rule *Tr*.

As in the proof of Lemma 5.7, only rule *Tr* can remove the new term v from the derivation. Consider an instance such that v does not appear anywhere below in the derivation:

$$\frac{c \leqslant v \quad v \leqslant d}{c \leqslant d} \; Tr \tag{1}$$

If the premiss $c \leqslant v$ is concluded by LM_1 or LM_2, then c is identical to v. The left premiss of *Tr* is $v \leqslant v$, but then the right premiss is identical to the conclusion and there is a loop. Rules RJ_1, RJ_2 cannot have as a conclusion $c \leqslant v$ or else v is in a *J*-atom. The remaining cases are those where $c \leqslant v$ has been concluded by *Tr*, *LJ*, or *RM*. With *Tr*, we permute up the *Tr* that removes v:

$$\frac{\dfrac{c \leqslant e \quad e \leqslant v}{c \leqslant v} \; Tr \quad v \leqslant d}{c \leqslant d} \; Tr \qquad \rightsquigarrow \qquad \frac{c \leqslant e \quad \dfrac{e \leqslant v \quad v \leqslant d}{e \leqslant d} \; Tr}{c \leqslant d} \; Tr \tag{2}$$

With *LJ*, there is some premiss of the form $J(e, f, c)$:

$$\frac{\dfrac{J(e, f, c) \quad e \leqslant v \quad f \leqslant v}{c \leqslant v} \; LJ \quad v \leqslant d}{c \leqslant d} \; Tr$$

Tr is permuted up as follows:

$$\frac{J(e, f, c) \quad \dfrac{e \leqslant v \quad v \leqslant d}{e \leqslant d} \; Tr \quad \dfrac{f \leqslant v \quad v \leqslant d}{f \leqslant d} \; Tr}{c \leqslant d} \; LJ \tag{3}$$

The permutation of *Tr* that removes v as in (2) and (3) is repeated until the left premiss has been concluded by *RM*. We then have some term c' such that

$$
\frac{\dfrac{M(a,b,v) \quad c' \leqslant a \quad c' \leqslant b}{c' \leqslant v}\, RM \quad v \leqslant d}{c' \leqslant d}\, Tr \tag{4}
$$

Now consider the right premiss $v \leqslant d$ of (4). Rules RJ_1, RJ_2, and LJ would give a *J*-atom with term v, so the possible rules are LM_1, LM_2, RM, and *Tr*. With *Tr* we permute similarly to (2):

$$
\frac{c' \leqslant v \quad \dfrac{v \leqslant g \quad g \leqslant d}{v \leqslant d}\, Tr}{c' \leqslant d}\, Tr \qquad \rightsquigarrow \qquad \frac{\dfrac{c' \leqslant v \quad v \leqslant g}{c' \leqslant g}\, Tr \quad g \leqslant d}{c' \leqslant d}\, Tr \tag{5}
$$

With *RM*, there is some premiss of the form $M(g, h, d)$:

$$
\frac{c' \leqslant v \quad \dfrac{M(g,h,d) \quad v \leqslant g \quad v \leqslant h}{v \leqslant d}\, RM}{c' \leqslant d}\, Tr
$$

Tr is permuted up as follows:

$$
\frac{M(g,h,d) \quad \dfrac{c' \leqslant v \quad v \leqslant g}{c' \leqslant g}\, Tr \quad \dfrac{c' \leqslant v \quad v \leqslant h}{c' \leqslant h}\, Tr}{c' \leqslant d}\, RM \tag{6}
$$

The permutation of *Tr* that removes v as in (5) and (6) is repeated until for some term d' an atom $v \leqslant d'$ has been concluded by LM_1 or LM_2. Then d' is identical to a or to b and step (4) has become one of:

$$
\frac{\dfrac{M(a,b,v) \quad c' \leqslant a \quad c' \leqslant b}{c' \leqslant v}\, RM \quad \dfrac{M(a,b,v)}{v \leqslant a}\, LM_1}{c' \leqslant a}\, Tr
$$

$$
\frac{\dfrac{M(a,b,v) \quad c' \leqslant a \quad c' \leqslant b}{c' \leqslant v}\, RM \quad \dfrac{M(a,b,v)}{v \leqslant b}\, LM_2}{c' \leqslant b}\, Tr
$$

Both derivations have a loop. Deletion of the part of the derivation between the two occurrences of the same formula deletes also the assumption $M(a, b, v)$. Thus, in the end there is no new term v in a transformed loop-free derivation, and therefore no instance of rule *EM*. The conclusion now follows by Lemma 5.7. QED.

Theorem 5.9 gives the main theorem of Skolem (1920) as a corollary:

Corollary 5.10. Conservativity of existential rules. *If an O-atom is derivable from given atoms in lattice theory, it is derivable without rules EM, EJ.*

Proof. By the subterm property, there cannot be eigenvariables in a minimal derivation of an O-atom.   QED.

(c) **Derivability of universal formulas.** We first reduce the derivability of arbitrary atoms to the derivability of O-atoms and then apply the subterm property to conclude Skolem's theorem on the decidability of universal formulas.

Lemma 5.11. *Derivability in* **ReLT** *of an M-atom $M(a, b, c)$ from assumptions Γ reduces to the derivability of two O-atoms, and the same for J-atoms.*

Proof. Let v be a fresh variable. We show that $M(a, b, c)$ is derivable from assumptions Γ if and only if $v \leqslant c$ and $c \leqslant v$ are derivable from $M(a, b, v)$ and Γ.

If $M(a, b, c)$ is derivable from Γ, we have

$$
\cfrac{M(a, b, v) \qquad \cfrac{\begin{array}{c} \Gamma \\ \vdots \\ M(a, b, c) \end{array}}{c \leqslant a}\,LM_1 \qquad \cfrac{\begin{array}{c} \Gamma \\ \vdots \\ M(a, b, c) \end{array}}{c \leqslant b}\,LM_2}{c \leqslant v}\,RM
$$

and

$$
\cfrac{M(a, b, c) \qquad \cfrac{M(a, b, v)}{v \leqslant a}\,LM_1 \qquad \cfrac{M(a, b, v)}{v \leqslant b}\,LM_2}{v \leqslant c}\,RM
$$

with $\Gamma \vdots$ above $M(a, b, c)$.

In the other direction, assuming $v \leqslant c$ and $c \leqslant v$ derivable from $M(a, b, v)$ and Γ, we have

$$
\cfrac{[M(a, b, v)] \qquad \cfrac{\cfrac{\begin{array}{c} [M(a, b, v)],\ \Gamma \\ \vdots \\ v \leqslant c \end{array} \qquad \begin{array}{c} [M(a, b, v)],\ \Gamma \\ \vdots \\ c \leqslant v \end{array}}{M(a, b, c)}\,SM}{M(a, b, c)}\,EM}{M(a, b, c)}
$$

Since v was chosen fresh, the variable restriction in rule $E\,M$ is met.   QED.

Theorem 5.12. Derivability of an atom from given atoms. *The derivability of an atom from given atoms in* **ReLT** *is decidable.*

Proof. By Lemma 5.11, we can assume the conclusion to be an O-atom. By the proof of Theorem 5.9, we can assume that there are no existential rules. By the subterm property, only a bounded number of terms need be used in instances of rules. Therefore the number of loop-free derivations that end with the atom to be derived is also bounded.   QED.

Skolem (1920, pp. 123–124) claims that the derivability of an arbitrary universal formula reduces to what Theorem 5.12 establishes. He uses the notion of conjunctive normal form for this, by which a universal formula can be expressed as a conjunction of disjunctions of the form

$$\neg P_1 \vee \ldots \vee \neg P_m \vee Q_1 \vee \ldots \vee Q_n.$$

We consider the equivalent implications of the form

$$P_1 \& \ldots \& P_m \supset Q_1 \vee \ldots \vee Q_n.$$

Skolem now states that at least one Q_i should follow under $P_1 \& \ldots \& P_m$, and this can be decided by Theorem 5.12, by going through the Q_i one by one. There is a general result about natural deduction and its extension by mathematical rules that explains why the method works:

Theorem 5.13. Disjunction property under atomic assumptions. *If $A \vee B$ is derivable from the atomic assumptions $P_1, \ldots, P_m$ in* **NI***, then one of A and B is derivable.*

Proof. Consider a normal derivation. If the last rule is an elimination, its major premiss remains an open assumption, but all the open assumptions are atomic. Therefore the last rule must be a disjunction introduction. By leaving it out, a derivation of A or B is found.   QED.

Observe that the result holds in particular for pure intuitionistic natural deduction, without any added rules.

If $A \vee B$ is a disjunction of atoms $Q_1 \vee Q_2$, a normal derivation of Q_1 or of Q_2 uses only the mathematical rules. By Theorem 5.12, the derivability of Q_1 and of Q_2 from $P_1, \ldots, P_m$ is decidable. A disjunction such as $Q_1 \vee \ldots \vee Q_n$ is a shorthand for a formula $(\ldots (Q_1 \vee Q_2) \vee \ldots) \vee Q_n$ with just binary disjunctions. Repeated application of Theorem 5.12 shows that also in this case, one of the Q_i is derivable if all assumptions are atomic.

Skolem apparently did not see the possibility of having a number of genuine cases $Q_1, \ldots, Q_n$ for which the disjunction property would not in general hold. It is the specific character of the lattice axioms, namely that they are all Harrop formulas, that makes the proof of Theorem 5.13 go through: with axioms that contain genuine disjunctions, mathematical rules cannot be fully separated from the logical rules of disjunction in extensions of natural deduction.

(d) Further decidable classes of formulas. The standard decidable classes of formulas of pure predicate calculus include the quantifier prefix classes $\forall \ldots \forall \exists \ldots \exists$ and $\forall \ldots \forall \exists \forall \ldots \forall$, their degenerate cases, etc. The formulation of lattice theory with existential axioms makes it a theory expressible in the language of pure predicate calculus, that is, without constants or functions. Consider those prefix classes that have a bounded Herbrand expansion. By the subterm property, proof search terminates for these classes, and the following result is obtained:

Theorem 5.14. Standard decidable classes. *Let QA be a formula in prenex form, with a quantifier prefix Q such that the corresponding Herbrand disjunction is bounded. Then derivability of QA in* **ReLT** *is decidable.*

Notes to Chapter 5

The extension of sequent calculus by rules that correspond to axioms that are geometric implications was developed systematically in Negri (2003). The contents of this chapter are an adaptation of geometric and co-geometric theories to natural deduction as a logical calculus. The class of co-geometric theories was identified first in Negri and von Plato (2005). The extension of natural deduction by geometric rules was used by Simpson (1994), as we found out after the extension of sequent calculus in Negri (2003) had been carried through.

Skolem's forgotten work in lattice theory was rediscovered in the 1990s (see Burris 1995 and Freese, Ježek, and Nation 1995). A review and reconstruction of Skolem's pioneering contributions to lattice theory is found in von Plato (2007).

The approach to relational lattice theory of Section 5.3 comes from Negri and von Plato (2004). It is different from Skolem's, though we did not realize that at the time of writing our paper in the spring of 2002. Skolem's proof uses, in fact, eigenvariables in the way of Section 4.4 above, as explained in detail in von Plato (2007).

PART II

Proof systems based on sequent calculus

6 | Rules of proof: sequent calculus

6.1 From natural deduction to sequent calculus

(a) Notation and rules for sequent calculus. Sequent calculus has a notation that displays, for each step of inference of a derivation, the open assumptions. Each formula C has the open assumptions Γ it depends on listed on
the same line:

$$\Gamma \to C$$

Sequent calculus can be regarded as a formal theory of the **derivability
relation** between a conclusion and the assumptions it depends on. In a
sequent $\Gamma \to C$, the left, assumption side Γ is called the **antecedent** and the
right, conclusion side C the **succedent**.

In Gentzen's original formulation of 1934–35, the assumptions Γ, Δ, Θ
were finite sequences, or **lists** as we would now say. Thus Gentzen had rules
that permitted the exchange of order of formulas in a sequence. We instead
consider assumptions **finite multisets**, that is, lists with multiplicity but no
order.

The rules of natural deduction show only the active formulas, and the
remaining open assumptions are left implicit. We can make the assumptions
explicit and write, for example, instead of

$$\frac{A \quad B}{A \& B} \,\&I$$

as follows:

$$\frac{\begin{array}{cc} \Gamma & \Delta \\ \vdots & \vdots \\ A & B \end{array}}{A \& B} \,\&I$$

Here the dots are an informal schematic notation for derivability from
assumptions.

When the derivability relation (the dots) is made formal, the introduction rules of natural deduction become the following **right rules** of sequent calculus, in which the comma represents multiset union:

Table 6.1 Right rules of sequent calculus

$$\frac{\Gamma \rightarrow A \quad \Delta \rightarrow B}{\Gamma, \Delta \rightarrow A\&B} \; R\& \qquad \frac{A, \Gamma \rightarrow B}{\Gamma \rightarrow A \supset B} \; R\supset \qquad \frac{\Gamma \rightarrow A}{\Gamma \rightarrow A \vee B} \; R\vee_1 \qquad \frac{\Gamma \rightarrow B}{\Gamma \rightarrow A \vee B} \; R\vee_2$$

The formula with the connective in a rule is the **principal** formula of that rule, and its components in the premisses the **active** formulas. The Greek letters denote possible additional assumptions that are not active in a rule; they are called the **contexts** of the rules.

The elimination rules of natural deduction correspond to the **left rules** of sequent calculus.

Table 6.2 Left rules of sequent calculus

$$\frac{A, B, \Gamma \rightarrow C}{A\&B, \Gamma \rightarrow C} \; L\& \qquad \frac{A, \Gamma \rightarrow C \quad B, \Delta \rightarrow C}{A \vee B, \Gamma, \Delta \rightarrow C} \; L\vee \qquad \frac{\Gamma \rightarrow A \quad B, \Delta \rightarrow C}{A \supset B, \Gamma, \Delta \rightarrow C} \; L\supset$$

In contrast to the rules of natural deduction, those of sequent calculus are **local**: the conclusion of a rule depends only on the premisses immediately above the inference line of the rule, unlike rule $\supset I$ and the E-rules of natural deduction.

We can ask whether the above rendering of the rules of natural deduction in another, local notation is sufficient for giving a system of sequent calculus. The answer is negative, for there are implicit rules of natural deduction that are neither introduction nor elimination rules, in particular:

1. The same formula can act as an assumption and as a conclusion in a derivation, as in

$$\frac{\overset{1}{[A]}}{A \supset A} \; \supset I, 1$$

This requires an addition to the logical rules of sequent calculus: the making of assumptions in natural deduction, as explained in the beginning of Section 2.1, corresponds to having **initial sequents** that have the form $A \to A$ and begin a derivation branch. The example derivation becomes:

$$\frac{A \to A}{\to A \supset A} \, R\supset$$

2. It is possible to close assumptions that have not been made, in a **vacuous discharge**, as in the following derivation:

$$\frac{\dfrac{\overset{1}{[A]}}{B \supset A} \, \supset I}{A \supset (B \supset A)} \, \supset I,1$$

3. It is possible to close several occurrences of the same formula at once, in a **multiple discharge**, as in the following derivation:

$$\frac{\dfrac{\dfrac{[A \supset (\overset{2}{A} \supset B)] \quad \overset{1}{[A]}}{A \supset B} \, \supset E \qquad \overset{1}{[A]}}{\dfrac{B}{A \supset B} \, \supset I,1}}{(A \supset (A \supset B)) \supset (A \supset B)} \, \supset I,2$$

4. It is possible to replace an assumption A in a derivation by a derivation of A and thus to obtain a derivation by **substitution**, or the **composition** of two derivations, C from Γ, A and A from Δ, as in

$$\begin{array}{c} \Delta \\ \vdots \\ \Gamma, \; A \\ \vdots \\ C \end{array}$$

The **structural rules** of sequent calculus correspond to the natural deduction construction principles 2–4 (sometimes also 1 is included).

Weakening introduces an extra assumption in the antecedent:

$$\frac{\Gamma \to C}{A, \Gamma \to C} \, Wk$$

In the notation of sequent calculus, example 2 becomes

$$\frac{\dfrac{\dfrac{A \to A}{A, B \to A} \, Wk}{A \to B \supset A} \, R\supset}{\to A \supset (B \supset A)} \, R\supset$$

It is seen that weakening corresponds to vacuous discharge in natural deduction.

Contraction is the rule:

$$\frac{A, A, \Gamma \rightarrow C}{A, \Gamma \rightarrow C} \, \mathit{Ctr}$$

Example 3 becomes the sequent calculus derivation

$$\cfrac{\cfrac{A \rightarrow A \quad \cfrac{\cfrac{A \rightarrow A \quad B \rightarrow B}{A \supset B, A \rightarrow B} \, L\supset}{A \supset (A \supset B), A, A \rightarrow B} \, L\supset}{\cfrac{A \supset (A \supset B), A \rightarrow B}{\cfrac{A \supset (A \supset B) \rightarrow A \supset B}{\rightarrow (A \supset (A \supset B)) \supset (A \supset B)} \, R\supset} \, R\supset} \, \mathit{Ctr}}{}$$

Here it is seen that contraction corresponds to a use of multiple discharge in natural deduction.

If assumptions are treated as sets instead of multisets, contraction is built into the system and is no longer expressed as a distinct rule. This innocent-looking change – often advocated in the literature as a simplification of the rules of sequent calculus – cannot be formalized in any simple way, as we shall see in a while.

Cut is the rule:

$$\frac{\Gamma \rightarrow A \quad A, \Delta \rightarrow C}{\Gamma, \Delta \rightarrow C} \, \mathit{Cut}$$

The use of cut corresponds in natural deduction to the composition of two derivations, but not only that: cut is needed in the translation from natural deduction to sequent calculus to express those instances of elimination rules in which the major premiss is derived, that is, not an assumption. These are the non-normal instances of elimination rules as in Definition 2.4.

Sometimes cut is explained through the familiar practice in mathematics of breaking proofs into lemmas, but it is essential that the cut formula in the right premiss be derived by a left rule at some stage. Use of the left rule corresponds to analysing the assumption into its components as in the E-rules of natural deduction. This would be the only way to put an assumption into proper use.

The propositional part of the sequent calculus obtained directly from natural deduction with general elimination rules will be called **G0i**. Its rules are now completely determined:

Table 6.3 The sequent calculus G0ip

Initial sequents

$A \to A$

Logical rules

$$\frac{A, B, \Gamma \to C}{A\&B, \Gamma \to C} \, L\& \qquad \frac{\Gamma \to A \quad \Delta \to B}{\Gamma, \Delta \to A\&B} \, R\&$$

$$\frac{A, \Gamma \to C \quad B, \Delta \to C}{A \vee B, \Gamma, \Delta \to C} \, L\vee \qquad \frac{\Gamma \to A}{\Gamma \to A \vee B} \, R\vee 1 \qquad \frac{\Gamma \to B}{\Gamma \to A \vee B} \, R\vee 2$$

$$\frac{\Gamma \to A \quad B, \Delta \to C}{A \supset B, \Gamma, \Delta \to C} \, L\supset \qquad \frac{A, \Gamma \to B}{\Gamma \to A \supset B} \, R\supset$$

$$\frac{}{\bot \to C} \, L\bot$$

Structural rules

$$\frac{\Gamma \to C}{A, \Gamma \to C} \, Wk \qquad \frac{A, A, \Gamma \to C}{A, \Gamma \to C} \, Ctr$$

Initial sequents are often called 'logical axioms' in the literature. The last sequent in a derivation is called its **endsequent**.

The rule of cut is not a part of the above calculus, because it is eliminable.

(b) 'Sequents as sets'. The suggestion has been made repeatedly of treating the antecedents of sequents $\Gamma \to C$ as sets of formulas, instead of lists or multisets. Let us look at rule $L\&$ of Table 6.3 in this light. Let the convention be that the curly brackets are left out in sets that form the antecedents, with A, B standing for the set $\{A, B\}$ and $A\&B$ for the set $\{A\&B\}$. Let Γ in rule $L\&$ be empty, and we have what seems like a perfectly good rule instance:

$$\frac{A, B \to C}{A\&B \to C} \, L\&$$

If it happened that $A \equiv B$, the antecedent would be the set $\{A\}$ and there is no instance of rule $L\&$ that would have as a conclusion the sequent $A\&A \to C$.

Rules of inference are part of a definition of the formal notion of a derivation. They convert derivations of their premises into a derivation of their conclusion. If it were possible to really treat 'contexts as sets', even syntactic identity, as in $A \equiv B$ in the above example, would not be needed to fault suggested instances of rules, but mere equality of objects. Whenever the notion of a set is put into use, a notion of equality of sets is also put into use. Two sets are equal if each member of one is a member of the other and *vice versa*. By the extensionality of sets, if $a \in S$ and $a = b$, then also $b \in S$. Thus, a further notion of equality of objects is involved when using sets. By the extensionality of sets, $\{a, b\} = \{a\}$ if $a = b$. In the rules of sequent calculus, the substitutability of equal sets would lead to syntactically incorrect instances of the logical rules.

To remedy the defect, one would have to say that the rules work on expressions for sets, not on sets directly, and to introduce a notation for the equality of expressions for sets, say $\Gamma \sim \Delta$. The axioms for such equality would include, next to reflexivity, symmetry, and transitivity, at least the following, with $n, m \geqslant 0$:

I $\{A_1, \ldots, A_n, A, B, B_1, \ldots, B_m\} \sim \{A_1, \ldots, A_n, B, A, B_1, \ldots, B_m\}$,

II $\{A_1, \ldots, A_n, A, A, B_1, \ldots, B_m\} \sim \{A_1, \ldots, A_n, A, B_1, \ldots, B_m\}$.

For example, given that we want to derive the sequent $B \& A \to C$ from the sequent $A, B \to C$, we would have to use the first axiom to conclude that $\{A, B\} \sim \{B, A\}$. When this step is made formal, the above two axioms have to be converted into rules. The first is symmetric in its left and right part, the second not, so we have three rules, called here E, C, and W, respectively:

Table 6.4 Structural rules for contexts as sets

$$\frac{A_1, \ldots, A_n, A, B, B_1, \ldots, B_m \to C}{A_1, \ldots, A_n, B, A, B_1, \ldots, B_m \to C} E \qquad \frac{A_1, \ldots, A_n, A, A, B_1, \ldots, B_m \to C}{A_1, \ldots, A_n, A, B_1, \ldots, B_m \to C} C$$

$$\frac{A_1, \ldots, A_n, A, B_1, \ldots, B_m \to C}{A_1, \ldots, A_n, A, A, B_1, \ldots, B_m \to C} W$$

The derivation of $B \& A \to C$ from $A, B \to C$ can now be made formal:

$$\frac{\dfrac{A, B \to C}{B, A \to C} E}{B \& A \to C} L\&$$

The derivation of $A \& A \to C$ from $A \to C$ is effected as follows:

$$\frac{\dfrac{A \to C}{A, A \to C} W}{A \& A \to C} L\&$$

Similar examples for rule C can be easily constructed. The formalization of the 'contexts as sets' idea leads to structural rules for the handling of equal contexts. In the above, rule E is just Gentzen's rule of exchange of the order of formulas in lists, rule C is the rule of contraction, and rule W a special case of the rule of weakening. Thus, nothing is gained but the structural rules just rendered informal by the 'contexts as sets' treatment.

(c) Desiderata on sequent calculi. Next we shall outline some desiderata on sequent calculi with a view to applications in proof search.

The rules of sequent calculus can have **independent** or **shared** contexts. The two styles in the right rule for conjunction give the rules

Table 6.5 Rules with independent and shared contexts

$$\frac{\Gamma \to A \quad \Delta \to B}{\Gamma, \Delta \to A\&B} \qquad \frac{\Gamma \to A \quad \Gamma \to B}{\Gamma \to A\&B}$$

Context-independent and context-sharing rules are easily seen to be equivalent in the presence of the structural rules, in the sense that each rule of one style is derivable from the corresponding rule of the other style. However, the two styles are not equivalent for proof search purposes. If the rules of sequent calculus are used to look for a derivation root first, from the endsequent to be derived, application of context-independent rules leads to an explosion of the combinatorial possibilities of splitting the context in the conclusion between the premisses of two-premiss rules. With context-sharing rules, the premisses are uniquely determined once the principal formula of the sequent to be derived is chosen.

Cut elimination is the best-known desired property of sequent calculus. If we look at the rules, all formulas in the premisses of each rule are found also in its conclusion, except for cut. Thus, we say that cut-free derivations have the **subformula property**, but in the presence of cut, the property is no longer guaranteed. Thus one of the main tasks of structural proof theory is the design of sequent calculi in which cut is an **eliminable** or **admissible** rule. Weakening is easily eliminated by letting initial sequents have the form $A, \Gamma \to A$ instead of $A \to A$. Contraction, instead, can be as 'bad' as cut, as concerns a root-first search for a derivation of a given sequent: formulas in antecedents can be multiplied with no end.

(d) Classical propositional logic. The rule obtained by a direct translation of the natural deduction rule of excluded middle of Chapter 2,

$$\frac{A, \Gamma \to C \quad \neg A, \Gamma \to C}{\Gamma \to C}$$

is not good for proof search purposes, because A is an arbitrary formula that *a priori* need have no relation to the formulas in Γ, C. Nevertheless, as mentioned in Chapter 2, the formula A can be restricted to atoms and indeed to atoms from C (cf. *Structural Proof Theory*, theorem 8.6.6). This works, however, only for the propositional part of classical logic, but not for the quantifiers.

(e) **Multisuccedent sequents.** Gentzen's formulation of classical logic is based on the extension of the notion of a sequent into a **multisuccedent** sequent. Sequents become expressions of the form

$$\Gamma \rightarrow \Delta$$

in which Γ and Δ are both multisets of formulas.

An intuitionistic system of sequent calculus is obtained as a special case of the classical system, by a restriction on the context in implication rules (and for predicate logic in the right rule for the universal quantifier as well). In this way we gain the advantage of a uniform formalism for intuitionistic and classical logic. Also Gentzen's original classical calculus, denoted **LK** in his first paper on the topic, was designed so that an intuitionistic calculus, denoted **LI**, was directly a special case of the classical calculus, one in which the succedent consisted of at most one formula. Thus, this calculus permitted sequents with empty antecedents (no assumptions in natural deduction), empty succedents (falsity $\perp$ in natural deduction), and even sequents that had both an empty antecedent and succedent.

In Gentzen (1934–35), what is sometimes called the **denotational** interpretation of multisuccedent sequents was given: a sequent $\Gamma \rightarrow \Delta$ expresses that the conjunction of the formulas in Γ implies the disjunction of the formulas in Δ.

The **operational** interpretation of single succedent sequents $\Gamma \rightarrow C$ through natural deduction is: from the assumptions Γ, the conclusion C **can be derived**. There is, however, no obvious way to extend it into sequents with multiple succedents.

Again in Gentzen (1938), the multisuccedent calculus is explained as the natural representation of the **division into cases** often found in mathematical proofs. Thus the antecedent Γ gives the **open assumptions** and the succedent Δ the **open cases** of a derivation.

The logical rules change and combine open assumptions and cases: $L\&$ replaces the open assumptions A, B by the open assumption $A\&B$; the dual multisuccedent rule $R\vee$ changes the open cases A, B into the open case

$A \vee B$, and so on. If there is just one case, we have an ordinary conclusion from open assumptions. The other limiting case with no formula in the succedent, as in $\Gamma \rightarrow$, corresponds to the empty case, that is, impossibility.

In a multisuccedent formulation of sequent calculus, the classical law of excluded middle is derivable as follows:

$$\frac{\dfrac{A \rightarrow A, \bot}{\rightarrow A, \neg A}}{\rightarrow A \vee \neg A}$$

(f) Sequent calculi with invertible rules. As noted, calculi with independent contexts are not suited for a root-first proof search, because the contexts in the conclusion could be divided in the premisses in different ways. With shared contexts, instead, the conclusion of a rule determines the premisses in a unique way. What is more, it happens with suitably formulated rules that the derivability between premisses and a conclusion goes both ways:

Definition 6.1. Invertibility. *A rule is* **invertible** *if from the derivability of a sequent of the form of its conclusion, the derivability of the corresponding premiss or premisses follows.*

For example, the single invertible rule

$$\frac{A, B, \Gamma \rightarrow \Delta}{A \& B, \Gamma \rightarrow \Delta}$$

is better in this respect than the two non-invertible rules

$$\frac{A, \Gamma \rightarrow \Delta}{A \& B, \Gamma \rightarrow \Delta} \qquad \frac{B, \Gamma \rightarrow \Delta}{A \& B, \Gamma \rightarrow \Delta}$$

If a premiss of one of the latter rules is given, say the first, it can be weakened into a premiss of the single rule by the addition of B in the antecedent, and then the single rule gives the conclusion. In the other direction, if a premiss of the single rule is given, the two rules can be applied, followed by a contraction on $A \& B$ to arrive at the conclusion of the single rule. Formally, we have:

$$\frac{\dfrac{A, \Gamma \rightarrow C}{A, B, \Gamma \rightarrow C}\,Wk}{A \& B, \Gamma \rightarrow C} \qquad \frac{\dfrac{\dfrac{A, B, \Gamma \rightarrow C}{A, A \& B, \Gamma \rightarrow C}}{A \& B, A \& B, \Gamma \rightarrow C}}{A \& B, \Gamma \rightarrow C}\,Ctr$$

Thus, the single rule is equivalent to the two rules. An example shows that the two rules cannot be invertible. The sequent $A \& B \rightarrow A \& B$ is derivable because it is an initial sequent. Neither of $A \to A\&B$ or $B \to A\&B$ is derivable, however.

Summing up, the desiderata for our sequent calculus are: multisuccedent, with context-sharing rules, admissible structural rules, and invertible logical rules. All of these desiderata are met in the following sequent calculus **G3c**.

Table 6.6 The sequent calculus G3c

Initial sequents

$P, \Gamma \to \Delta, P$

Logical rules

$$\frac{A, B, \Gamma \to \Delta}{A\&B, \Gamma \to \Delta} \; L\& \qquad\qquad \frac{\Gamma \to \Delta, A \quad \Gamma \to \Delta, B}{\Gamma \to \Delta, A\&B} \; R\&$$

$$\frac{A, \Gamma \to \Delta \quad B, \Gamma \to \Delta}{A \vee B, \Gamma \to \Delta} \; L\vee \qquad\qquad \frac{\Gamma \to \Delta, A, B}{\Gamma \to \Delta, A \vee B} \; R\vee$$

$$\frac{\Gamma \to \Delta, A \quad B, \Gamma \to \Delta}{A \supset B, \Gamma \to \Delta} \; L\supset \qquad\qquad \frac{A, \Gamma \to \Delta, B}{\Gamma \to \Delta, A \supset B} \; R\supset$$

$$\frac{}{\bot, \Gamma \to \Delta} \; L\bot$$

Observe that initial sequents have an atomic formula on both sides of the arrow.

An intuitionistic system is obtained as a special case of **G3c** by a modification of the implication rules:

Table 6.7 Implication rules for the intuitionistic calculus G3im

---

$$\frac{A \supset B, \Gamma \to A \quad B, \Gamma \to \Delta}{A \supset B, \Gamma \to \Delta} \; L\supset \qquad\qquad \frac{A, \Gamma \to B}{\Gamma \to \Delta, A \supset B} \; R\supset$$

In **G3im**, the letter **m** stands for multisuccedent. The principal formula of rule $L\supset$ is repeated in the antecedent of the left premiss. In rule $R\supset$, there is no context in the succedent of the premiss. The reasons for these complications are explained in *Structural Proof Theory*, section 5.3.

We use in the results that follow the notation $\vdash_n \Gamma \to \Delta$ to indicate that the sequent $\Gamma \to \Delta$ has a derivation of height bounded by n, where the height of a derivation is its height as a tree.

Theorem 6.2. Height-preserving inversion. *All rules of* **G3c** *are invertible, with height-preserving inversion.*

Proof. We have to show that if $\vdash_n \Gamma \rightarrow \Delta, A\&B$, then $\vdash_n \Gamma \rightarrow \Delta, A$ and $\vdash_n \Gamma \rightarrow \Delta, B$, and analogously for the other connectives. The proof is by induction on n.

If $\Gamma \rightarrow \Delta, A\&B$ is an initial sequent or conclusion of $L\bot$, then, $A\&B$ not being atomic, also $\Gamma \rightarrow \Delta, A$ and $\Gamma \rightarrow \Delta, B$ are initial sequents or conclusions of $L\bot$. Assume height-preserving inversion up to height n, and let $\vdash_{n+1} \Gamma \rightarrow \Delta, A\&B$. There are two cases.

If $A\&B$ is not principal in the last rule, it has one or two premises $\Gamma' \rightarrow \Delta', A\&B$ and $\Gamma'' \rightarrow \Delta'', A\&B$, of derivation height $\leqslant n$, so by inductive hypothesis, $\vdash_n \Gamma' \rightarrow \Delta', A$ and $\vdash_n \Gamma' \rightarrow \Delta', B$ and $\vdash_n \Gamma'' \rightarrow \Delta'', A$ and $\vdash_n \Gamma'' \rightarrow \Delta'', B$. Now apply the last rule to these premises to conclude $\Gamma \rightarrow \Delta, A$ and $\Gamma \rightarrow \Delta, B$ with a height of derivation $\leqslant n+1$. If $A\&B$ is principal in the last rule, the premises $\Gamma \rightarrow \Delta, A$ and $\Gamma \rightarrow \Delta, B$ have derivations of height $\leqslant n$.   QED.

Next we consider the structural rules:

Table 6.8 The rules of weakening and contraction

$$\frac{\Gamma \rightarrow \Delta}{A, \Gamma \rightarrow \Delta} LW \qquad \frac{\Gamma \rightarrow \Delta}{\Gamma \rightarrow \Delta, A} RW \qquad \frac{A, A, \Gamma \rightarrow \Delta}{A, \Gamma \rightarrow \Delta} LC \qquad \frac{\Gamma \rightarrow \Delta, A, A}{\Gamma \rightarrow \Delta, A} RC$$

These rules are all admissible in the **G3**-calculi:

Theorem 6.3. Height-preserving admissibility of weakening

> *If* $\vdash_n \Gamma \rightarrow \Delta$, *then* $\vdash_n A, \Gamma \rightarrow \Delta$.
> *If* $\vdash_n \Gamma \rightarrow \Delta$, *then* $\vdash_n \Gamma \rightarrow \Delta, A$.

Proof. Assume $\Gamma \rightarrow \Delta$ to have been derived with height n. Add the formula A to each antecedent in the derivation, to get a derivation of $A, \Gamma \rightarrow \Delta$ with height n. The proof is similar for right weakening.   QED.

Theorem 6.4. Height-preserving admissibility of contraction

> *If* $\vdash_n A, A, \Gamma \rightarrow \Delta$, *then* $\vdash_n A, \Gamma \rightarrow \Delta$.
> *If* $\vdash_n \Gamma \rightarrow \Delta, A, A$, *then* $\vdash_n \Gamma \rightarrow \Delta, A$.

Proof. The proof is by a simultaneous induction on the height of the derivation for left and right contraction, using height-preserving invertibility of the rules. The details are found in *Structural Proof Theory,* theorem 3.2.2.   QED.

Theorem 6.5. *The rule of cut,*

$$\frac{\Gamma \rightarrow \Delta, D \quad D, \Gamma' \rightarrow \Delta'}{\Gamma, \Gamma' \rightarrow \Delta, \Delta'} \; Cut$$

is admissible in **G3c**.

Proof. The proof is just sketched here. The principle is to consider an uppermost cut in a derivation. As long as the cut formula is not principal in the right premiss of cut, it is permuted up. When it is principal in the right but not in the left premiss, it is permuted up at left. This step corresponds to a permutative conversion in natural deduction. Finally, we come to the essential case in which cut is principal in both premisses. In the case of conjunction, we have the part of the derivation

$$\frac{\dfrac{\Gamma \rightarrow \Delta, A \quad \Gamma \rightarrow \Delta, B}{\Gamma \rightarrow \Delta, A\&B} \; R\& \quad \dfrac{A, B, \Gamma' \rightarrow \Delta'}{A\&B, \Gamma' \rightarrow \Delta'} \; L\&}{\Gamma, \Gamma' \rightarrow \Delta, \Delta'} \; Cut$$

This part of the derivation is transformed into

$$\frac{\Gamma \rightarrow \Delta, B \quad \dfrac{\Gamma \rightarrow \Delta, A \quad A, B, \Gamma' \rightarrow \Delta'}{B, \Gamma, \Gamma' \rightarrow \Delta, \Delta'} \; Cut}{\Gamma, \Gamma, \Gamma' \rightarrow \Delta, \Delta, \Delta'} \; Cut$$

Now there are two cuts on shorter formulas. Other connectives are treated similarly. Cuts are permuted up until one premiss of cut is an initial sequent. Then the conclusion of cut is the same as the other premiss of cut, modulo weakening, or else it is an initial sequent. For the details, see *Structural Proof Theory,* p. 54.   QED.

The proof of cut elimination for **G3c** shows a first advantage of the use of a contraction-free calculus: there is no need of **multicut**, a rule in which several occurrences of the cut formula are deleted in the cut. It was used by Gentzen to deal with a case of non-reducibility in the proof of cut elimination, the one in which one of the premisses of cut is derived by a contraction. Cut is a special case of multicut, so the elimination of multicut gives cut elimination as a corollary. However, by a deeper analysis, not just of the last rule applied in the premisses of cut, it is possible to prove cut elimination without using multicut even in the presence of the other structural rules.

Corollary 6.6. Subformula property. *Each formula in the derivation of a sequent* $\Gamma \rightarrow \Delta$ *in* **G3c** *is a subformula of* Γ, Δ.

As an immediate consequence, we have:

Corollary 6.7. Consistency. *The sequent* $\rightarrow$ *is not derivable.*

By the admissibility of weakening, if $\rightarrow$ were derivable, then also $\rightarrow \perp$ would be derivable. The converse is obtained by applying cut to $\rightarrow \perp$ and $\perp \rightarrow$, thus, an empty succedent behaves like $\perp$.

(g) Rules for the quantifiers. For the sake of completeness of presentation, we list here the quantifier rules of sequent calculus:

Table 6.9 The quantifier rules in sequent calculus **G3c**

$$\frac{A(t/x), \forall xA, \Gamma \rightarrow \Delta}{\forall xA, \Gamma \rightarrow \Delta} \, L\forall \qquad \frac{\Gamma \rightarrow \Delta, A(y/x)}{\Gamma \rightarrow \Delta, \forall xA} \, R\forall$$

$$\frac{A(y/x), \Gamma \rightarrow \Delta}{\exists xA, \Gamma \rightarrow \Delta} \, L\exists \qquad \frac{\Gamma \rightarrow \Delta, \exists xA, A(t/x)}{\Gamma \rightarrow \Delta, \exists xA} \, R\exists$$

Rules $R\forall$ and $L\exists$ have an eigenvariable y, not free in the conclusion of the rules. Rules $L\forall$ and $R\exists$ have an arbitrary term t in the premiss. They have also a repetition of the principal formula in the premiss to guarantee the height-preserving admissibility of contraction.

The rules of Table 6.9 are those of the classical sequent calculus **G3c**. The quantifier rules of the intuitionistic multisuccedent calculus **G3im** are the same except that the context Δ in the premiss of rule $R\forall$, but not in its conclusion, must be empty. The reasons for this condition are as those for the analogous restriction in rule $R\supset$. The proofs of admissibility of the structural rules go through also when the quantifier rules are added.

6.2 Extensions of sequent calculus

(a) Cut elimination in the presence of axioms. It is well known that cut elimination fails in the presence of proper axioms. A simple counterexample is given in Girard (1987, p. 125). Let the axioms have the forms $A \supset B$ and

A. These are represented by the 'axiomatic sequents' $\rightarrow A \supset B$ and $\rightarrow A$. The sequent $\rightarrow B$ is derived from these axiomatic sequents, as in:

$$\cfrac{\rightarrow A \qquad \cfrac{\rightarrow A \supset B \qquad \cfrac{A \rightarrow A \qquad B \rightarrow B}{A, A \supset B \rightarrow B} L\supset}{A \rightarrow B} Cut}{\rightarrow B} Cut$$

However, there is no cut-free derivation of $\rightarrow B$.

Observe that if the axioms are converted into the rules

$$\cfrac{B \rightarrow C}{A \rightarrow C} \qquad \cfrac{A \rightarrow C}{\rightarrow C}$$

then the sequent $\rightarrow B$ has instead a cut-free derivation

$$\cfrac{\cfrac{B \rightarrow B}{A \rightarrow B}}{\rightarrow B}$$

The above example shows only the idea of the conversion of axioms into rules within a sequent calculus. In order to make the idea precise we have to look carefully at the proof of admissibility of the structural rules for the **G3** sequent calculi. This inspection shows how to convert axioms into rules while maintaining the admissibility of structural rules in the extended systems.

First, the rules that correspond to the Hilbert-style axioms have to be 'logic free'. The logical content of the mathematical axioms is absorbed into the combinatorics of the sequent calculus mathematical rules. Only atomic formulas should appear as active and principal in such rules.

Secondly, it is seen that the two rules act only on the antecedent side of sequents. The rules will have an arbitrary multiset in the succedent.

The most general scheme that corresponds to these principles is:

Table 6.10 The scheme of mathematical rules

$$\cfrac{Q_1, \Gamma \rightarrow \Delta \quad \ldots \quad Q_n, \Gamma \rightarrow \Delta}{P_1, \ldots, P_m, \Gamma, \rightarrow \Delta} R$$

Here Γ, Δ are arbitrary multisets; $P_1, \ldots, P_m$ in the conclusion the **principal atoms** in the rule; and $Q_1, \ldots, Q_n$ in the premisses the **active atoms** in the rule, with $m, n \geqslant 0$. In particular, a rule can have zero premisses.

The full rule scheme corresponds to the formula

$$P_1 \& \ldots \& P_m \supset Q_1 \vee \ldots \vee Q_n.$$

To see better what forms of axioms the rule scheme covers, we write out a few cases, together with their corresponding axiomatic statements in a Hilbert-style calculus. The rules for axioms of the forms $Q\&R$, $Q \lor R$, and $P \supset Q$ are, with two rules for $Q\&R$,

$$\frac{Q, \Gamma \to \Delta}{\Gamma \to \Delta} \, , \, \frac{R, \Gamma \to \Delta}{\Gamma \to \Delta} \qquad \frac{Q, \Gamma \to \Delta \quad R, \Gamma \to \Delta}{\Gamma \to \Delta} \qquad \frac{Q, \Gamma \to \Delta}{P, \Gamma \to \Delta}$$

The rules for axioms of the forms Q, $\neg P$, and $\neg(P_1\&P_2)$ are:

$$\frac{Q, \Gamma \to \Delta}{\Gamma \to \Delta} \qquad \frac{}{P, \Gamma \to \Delta} \qquad \frac{}{P_1, P_2, \Gamma \to \Delta}$$

In order to deal with the admissibility of contraction, we have to augment the rule scheme. Right contraction is unproblematic because of the arbitrary context Δ in the succedents of the rule scheme. To handle left contraction, assume there to be a derivation of $A, A, \Gamma \to \Delta$, and assume that the last rule is a mathematical one. Then the derivation of $A, A, \Gamma \to \Delta$ can be of three different forms: first, neither occurrence of A is principal in the rule; second, one is principal; third, both are principal. The first case is handled by a straightforward induction, and the second case by the method, familiar from the work of Kleene (1952) and exemplified by rule $L\supset$ of the calculus **G3im**, of repeating the principal formulas of the conclusion in the premises. Thus, the general rule scheme becomes

Table 6.11 The rule scheme with repetition of principal formulas

$$\frac{P_1, \ldots, P_m, Q_1, \Gamma \to \Delta \quad \ldots \quad P_1, \ldots, P_m, Q_n, \Gamma \to \Delta}{P_1, \ldots, P_m, \Gamma \to \Delta} R$$

Repetitions in the premises will make left contractions commute with rules that follow the scheme. For the remaining case, with both occurrences of formula A principal in the last rule, consider the situation with a Hilbert-style axiomatization. We have some axiom, say $\neg(a < b \,\&\, b < a)$ in the theory of strict linear order, and substitution of b with a produces $\neg(a < a \,\&\, a < a)$ that we routinely abbreviate to $\neg a < a$, irreflexivity of strict linear order. This step is in fact a contraction. For systems with rules, the case in which a substitution produces two identical formulas that are both principal in a mathematical rule is taken care of by the following:

Closure condition. *Given a system with mathematical rules, if it has a rule with a substitution instance of the form*

$$\frac{P_1, \ldots, P_{m-2}, P, P, Q_1, \Gamma \to \Delta \quad \ldots \quad P_1, \ldots, P_{m-2}, P, P, Q_n, \Gamma \to \Delta}{P_1, \ldots, P_{m-2}, P, P, \Gamma \to \Delta}$$

then the system has to contain also the rule

$$\frac{P_1, \ldots, P_{m-2}, P, Q_1, \Gamma \to \Delta \quad \ldots \quad P_1, \ldots, P_{m-2}, P, Q_n, \Gamma \to \Delta}{P_1, \ldots, P_{m-2}, P, \Gamma \to \Delta}$$

The condition is unproblematic, because the number of rules to be added to a given system of mathematical rules is bounded. Often the closure condition is superfluous, because the contracted rule is already a rule of the system.

Which axioms are representable as rules that follow the rule scheme? For classical systems, the answer is unproblematic. All universal axioms can have their propositional matrix converted to a conjunction of disjunctions of atoms and negations of atoms. Each conjunct can be converted into the classically equivalent form $P_1 \& \ldots \& P_m \supset Q_1 \vee \ldots \vee Q_n$ that is representable as a rule of inference. We therefore have

Proposition 6.8. *All classical quantifier-free axioms can be represented by rules that follow the rule scheme.*

The conversion to conjunctive normal form does not hold in general in intuitionistic logic, so we have for intuitionistic systems a smaller class of axioms convertible into rules that follow the rule scheme. See section 6.1(a) of *Structural Proof Theory* for details.

The following result, proved in detail in section 6.2 of *Structural Proof Theory*, holds for extensions of the **G3** sequent systems. We shall denote by **G3c*** (**G3im***) any extension of **G3** (**G3im**) by rules that follow the rule scheme and satisfy the closure condition:

Theorem 6.9. *All the structural rules (weakening, contraction, and cut) are admissible in* **G3c*** *and in* **G3im***. *Weakening and contraction are height-preserving admissible.*

Proof. For left weakening, since the two initial sequents and all rules have an arbitrary context in the antecedent, addition of the weakening formula to the antecedent of each sequent will give a derivation of $A, \Gamma \to \Delta$. For right weakening, addition of the weakening formula to the succedents of all sequents that are not followed by an instance of rule $R\supset$ or rule $R\forall$ gives a derivation of $\Gamma \to \Delta, A$.

The proof of admissibility of the contraction rules and of the cut rule for **G3im** requires the use of inversion lemmas. We observe that all the inversion lemmas that hold for **G3im** hold for **G3im***** as well. This is achieved by having only atomic formulas as principal in mathematical rules. For left contraction, the proof is by induction on the height of derivation of the premiss: if the premiss of contraction is an initial sequent or conclusion of rule $L\perp$, the conclusion also is an initial sequent or conclusion of $L\perp$.

If A is not principal in the last rule, either logical or mathematical, apply the inductive hypothesis to the premisses and then the rule.

If A is principal and the last rule is logical, for $L\&$ and $L\vee$ apply height-preserving invertibility, the inductive hypothesis, and then the rule. For $L\supset$ apply the inductive hypothesis to the left premiss, invertibility and the inductive hypothesis to the right premiss, and then the rule. If the last rule is $L\vee$, apply the inductive hypothesis to its premiss, and then $L\vee$. If the last rule is $L\exists$, apply height preserving invertibility of $L\exists$, the inductive hypothesis, and $L\exists$.

If the last rule is mathematical, A is an atomic formula P and there are two cases. In the first case one occurrence of A belongs to the context, another is principal in the rule, say $A \equiv P_m \equiv P$. The derivation ends with

$$\frac{Q_1, P_1, \ldots, P_{m-1}, P, P, \Gamma' \to \Delta \quad \ldots \quad Q_n, P_1, \ldots, P_{m-1}, P, P, \Gamma' \to \Delta}{P_1, \ldots, P_{m-1}, P, P, \Gamma' \to \Delta} R$$

and we obtain

$$\frac{\dfrac{Q_1, P_1, \ldots, P_{m-1}, P, P, \Gamma' \to \Delta}{Q_1, P_1, \ldots, P_{m-1}, P, \Gamma' \to \Delta} \, Ind \quad \ldots \quad \dfrac{Q_n, P_1, \ldots, P_{m-1}, P, P, \Gamma' \to \Delta}{Q_n, P_1, \ldots, P_{m-1}, P, \Gamma' \to \Delta} \, Ind}{P_1, \ldots, P_{m-1}, P, \Gamma' \to \Delta} R$$

In the second case both occurrences of A are principal in the rule, say $A \equiv P_{m-1} \equiv P_m \equiv P$; thus the derivation ends with

$$\frac{Q_1, P_1, \ldots, P_{m-2}, P, P, \Gamma' \to \Delta \quad \ldots \quad Q_n, P_1, \ldots, P_{m-2}, P, P, \Gamma' \to \Delta}{P_1, \ldots, P_{m-2}, P, P, \Gamma' \to \Delta} R$$

and we obtain

$$\frac{\dfrac{Q_1, P_1, \ldots, P_{m-2}, P, P, \Gamma' \to \Delta}{Q_1, P_1, \ldots, P_{m-2}, P, \Gamma' \to \Delta} \, Ind \quad \ldots \quad \dfrac{Q_n, P_1, \ldots, P_{m-2}, P, P, \Gamma' \to \Delta}{Q_n, P_1, \ldots, P_{m-2}, P, \Gamma' \to \Delta} \, Ind}{P_1, \ldots, P_{m-2}, P, \Gamma' \to \Delta} R$$

with the last rule given by the closure condition.

The proof of admissibility of right contraction in **G3im***** does not present any additional difficulty with respect to the proof of admissibility in **G3im** since in mathematical rules the succedent in both the premisses and the conclusion is an arbitrary multiset Δ. So in the case where the last rule in a derivation of $\Gamma \to \Delta, A, A$ is a mathematical rule, one simply proceeds by applying the inductive hypothesis to the premisses, and then the rule.

The rule of cut is proved admissible by induction on the length of A with subinduction on the sum of the heights of the derivations of the premisses of cut $\Gamma \to \Delta, A$ and $A, \Gamma' \to \Delta'$. We consider here in detail only the cases that arise from the addition of mathematical rules. The other cases are treated in the corresponding proof for the purely logical calculi **G3c** and **G3im**.

1. If the left premiss is the conclusion of a zero-premiss mathematical rule, then also the conclusion is, because such rules have an arbitrary context as succedent.

2. If the right premiss is the conclusion of a zero-premiss mathematical rule with A not principal in it, then also the conclusion is, for the same reason as in 1.

3. If the right premiss is the conclusion of a zero-premiss mathematical rule with A principal in it, A is atomic and we consider the left premiss. The case in which it is the conclusion of a zero-premiss mathematical rule is covered by 1. If it is an initial sequent with A not principal, the conclusion is an initial sequent; otherwise, Γ contains the atom A and the conclusion follows from the right premiss by weakening.

In the remaining cases we consider the last rule in the derivation of $\Gamma \to \Delta, A$. Since A is atomic, A is not principal in the rule. Let us consider the case of a mathematical rule. We transform the derivation, where P stands for $P_1, \ldots, P_m$,

$$\frac{\dfrac{Q_1, P, \Gamma'' \to \Delta, A \ \ldots \ Q_n, P, \Gamma'' \to \Delta, A}{P, \Gamma'' \to \Delta, A} R \qquad A, \Gamma' \to \Delta'}{P, \Gamma', \Gamma'' \to \Delta, \Delta'} Cut$$

into

$$\frac{\dfrac{Q_1, P, \Gamma'' \to \Delta, A \quad A, \Gamma' \to \Delta'}{Q_1, P, \Gamma', \Gamma'' \to \Delta, \Delta'} Cut \ \ldots \ \dfrac{Q_n, P, \Gamma'' \to \Delta, A \quad A, \Gamma' \to \Delta'}{Q_n, P, \Gamma', \Gamma'' \to \Delta, \Delta'} Cut}{P, \Gamma', \Gamma'' \to \Delta, \Delta'} R$$

where the cut has been replaced by n cuts with left premiss with derivation of lower height.

Logical rules are dealt with similarly, except for $R \supset$ and $R \forall$ in **G3im** that are handled in 4.

Next we have the cases in which both premisses are derived by rules that have premisses.

4. *A* is not principal in the left premiss. For rules other than $R\supset$ and $R\forall$ in **G3im**, the cut is permuted up to the premisses of the rule by which the left premiss was derived, with variable renaming to match the variable restrictions in the cases of quantifier rules. Rules $R\supset$ and $R\forall$ of **G3im** have a restriction by which *A* does not appear in the premiss, but the conclusion is obtained without cut by $R\supset$ or $R\forall$ and weakening.

5. *A* is principal in only the left premiss. Then *A* has to be a compound formula. Therefore, if the last rule of the right premiss is a mathematical rule, *A* cannot be principal in the rule, because only atomic formulas are principal in mathematical rules. In this case cut is permuted to the premisses of the right premiss of cut. If the rule is a logical one with *A* not principal in it, the usual reductions are applied.

6. *A* is principal in both premisses. This case can involve only logical rules and is dealt with as in the usual proof for pure logic.   QED.

In systems with mathematical rules we have a **weak subformula property**:

Theorem 6.10. *If* $\Gamma \rightarrow \Delta$ *is derivable in* **G3im*** *or* **G3c***,* *then all formulas in the derivation are either subformulas of the endsequent or atomic formulas.*

A priori, atomic formulas can be lost track of. However, it turns out in many cases that the weak subformula property is sufficient for a structural analysis of proofs. In the applications to the theories of order, lattice theory, and geometry that will be presented in the following chapters, we shall establish a subterm property for sequent calculus derivations, analogous to natural deduction, by which proof search can be resricted to known terms. Under this property, there is only a bounded number of distinct atomic formulas that are not visible from the endsequent to be derived.

A simple test for consistency for theories convertible to rules can be made by analyzing the possible derivations of $\rightarrow \perp$ in **G3c*** or **G3im***:

Theorem 6.11. *Assume that a theory convertible to rules is inconsistent. Then*

(i) *All rules in the derivation of* $\rightarrow \perp$ *are mathematical,*
(ii) *All sequents in the derivation have* $\perp$ *as a succedent,*
(iii) *Each branch in the derivation begins with a rule of the form*

$$\frac{}{P_1, \ldots, P_m \rightarrow \perp}$$

(iv) *The last step in the derivation is a rule of the form*

$$\frac{Q_1 \rightarrow \bot \quad \ldots \quad Q_n \rightarrow \bot}{\rightarrow \bot}$$

A detailed proof is given in *Structural Proof Theory*, theorem 6.4.2. Given a system of axioms, the conjuncts of the axioms in conjunctive normal form can be written in the form $P_1 \& \ldots \& P_m \supset Q_1 \vee \ldots \vee Q_n$, with $n, m \geqslant 0$. It follows that if an axiom system is inconsistent, there must be at least one such formula with $m = 0$, and another with $n = 0$. In other words, at least one formula is a negation and at least one an atom or a disjunction of atoms. Therefore, if among the axioms of a system there are neither atoms nor disjunctions of atoms, the axioms are consistent, and similarly if there are no negations.

(b) Four approaches to extension by axioms. The method presented in the previous section is not the only way to extend sequent calculi for the treatment of axiomatic theories. Here we list four different approaches and how they behave with respect to cut elimination and proof analysis. As a basis, we take the intuitionistic calculus **G3im**, and then we have: 1. Addition of axioms A into **G3im** in the form of sequents $\rightarrow A$ by which derivations can start. As is shown in the example at the beginning of this section, the method leads to a failure of cut elimination. 2. Gentzen's method in his (1938, sec. 1.4). Add 'mathematical groundsequents' of the form $P_1, \ldots, P_m \rightarrow Q_1, \ldots, Q_n$. By Gentzen's 'Hauptsatz', the cut rule can be permuted into such groundsequents and arbitrary cuts reduced to cuts on atoms. Weakening and contraction have to be added as explicit rules. 3. The method of Gentzen's consistency proof of elementary arithmetic without induction (1934–35, sec. IV.3). Treat axioms as a context Γ and prove results of the form $\Gamma \rightarrow \Delta$. Cut elimination applies but the resulting system is not contraction free. Arbitrary instances of the axioms may appear in the antecedent. 4. Axioms are converted into rules.

All these approaches are equivalent, but the fourth is the one best suited for the purposes of proof analysis. We define the four approaches more precisely as follows:

Definition 6.12

(a) *An extension of* **G3im** *through* **axioms** *is obtained by adding the structural rules and letting axiomatic sequents* $\rightarrow A$, *with A an axiom, begin derivations.*

(b) *An extension of* **G3im** *through* **groundsequents** *is obtained by adding the structural rules and letting sequents* $P_1, \ldots, P_m \to Q_1, \ldots, Q_n$, *corresponding to the axioms, begin derivations.*

(c) *An extension of* **G3im** *through an* **axiomatic context** *is obtained by letting instances of the axioms appear in the contexts of sequents.*

(d) *An extension of* **G3im** *through* **rules** *is obtained by the conversion of axioms into mathematical rules.*

Theorem 6.13. *The four types of extensions of sequent calculus are equivalent with respect to the derivability of sequents.*

Proof. Axiomatic sequents and groundsequents are interderivable by cuts. We show the equivalence of systems with rules to systems with axiomatic sequents and with axiomatic contexts. For transparency, we consider the formula $P \supset Q \vee R$ (*Split*), as other formulas convertible to rules are special cases or inessential generalizations of it.

1. *Equivalence of rules and axiomatic sequents:* The rule

$$\frac{Q, P \to \Delta \quad R, P \to \Delta}{P \to \Delta} \; Split$$

can be derived through the axiom $\to P \supset Q \vee R$ by means of cuts and contractions:

$$\cfrac{\to P \supset Q \vee R \quad \cfrac{\cfrac{P \supset Q \vee R, P \to P \quad Q \vee R, P \to Q \vee R}{P \supset Q \vee R, P \to Q \vee R} \; L\supset \quad \cfrac{Q, P \to \Delta \quad R, P \to \Delta}{Q \vee R, P \to \Delta} \; L\vee}{P \supset Q \vee R, P, P \to \Delta} \; Cut}{\cfrac{P, P \to \Delta}{P \to \Delta} \; LC} \; Cut$$

In the other direction, $\to P \supset Q \vee R$ is derivable by rule *Split*:

$$\cfrac{\cfrac{\cfrac{Q, P \to Q, R}{Q, P \to Q \vee R} \; R\vee \quad \cfrac{R, P \to Q, R}{R, P \to Q \vee R} \; R\vee}{P \to Q \vee R} \; Split}{\to P \supset Q \vee R} \; R\supset$$

2. *Equivalence of axiomatic contexts and rules:* Assume that $\Gamma \to \Delta$ was derived by rule *Split*, and show that $P \supset Q \vee R, \Gamma \to \Delta$ can be derived by the rules of **G3im**. We assume that *Split* is the last rule in the derivation, and therefore $\Gamma \equiv P, \Gamma'$. The premises of the rule are $Q, P, \Gamma' \to \Delta$ and $R, P, \Gamma' \to \Delta$; thus there are by the inductive hypothesis instances $A_1, \ldots, A_m$ and $A'_1, \ldots, A'_n$ of the axioms in the contexts such that

$$Q, P, \Gamma', A_1, \ldots, A_m \to \Delta \quad \text{and} \quad R, P, \Gamma', A'_1, \ldots, A'_n \to \Delta$$

are derivable in **G3im**. Structural rules can be used, and we have, in **G3im**, a derivation that starts with a weakening of the A_i and A'_j into a common context $A''_1, \ldots, A''_k$ of instances of axioms:

$$\cfrac{P \supset Q \vee R, P, \Gamma', A''_1, \ldots, A''_k \to P \qquad \cfrac{\cfrac{Q, P, \Gamma', A_1, \ldots, A_m \to \Delta}{Q, P, \Gamma', A''_1, \ldots, A''_k \to \Delta} LW \qquad \cfrac{R, P, \Gamma', A'_1, \ldots, A'_n \to \Delta}{R, P, \Gamma', A''_1, \ldots, A''_k \to \Delta} LW}{\cfrac{Q \vee R, P, \Gamma', A''_1, \ldots, A''_k \to \Delta}{P \supset Q \vee R, P, \Gamma', A''_1, \ldots, A''_k \to \Delta} L\supset} L\vee}{P \supset Q \vee R, P, \Gamma', A''_1, \ldots, A''_k \to \Delta}$$

The split formula and the $A''_1, \ldots, A''_k$ are instances of axioms.

In the other direction, let $\Gamma \to \Delta$ be derivable when axiom instances can be added to Γ. Suppose for simplicity that only the split axiom occurs in the context, i.e., that $P \supset Q \vee R, \Gamma \to \Delta$ is derivable. We have the derivation by rule *Split*

$$\cfrac{\cfrac{\cfrac{\cfrac{Q, P \to Q, R}{Q, P \to Q \vee R} R\vee \qquad \cfrac{R, P \to Q, R}{R, P \to Q \vee R} R\vee}{P \to Q \vee R} Split}{\to P \supset Q \vee R} R\supset \qquad P \supset Q \vee R, \Gamma \to \Delta}{\Gamma \to \Delta} Cut$$

By the admissibility of cut in **G3im***, the conclusion follows.   QED.

In systems with axioms and with groundsequents, cuts cannot be eliminated, whereas systems with axiomatic contexts and with rules are cut free. The strength of systems with rules is that they permit proofs by induction on rules used in a derivation, with some surprisingly simple, purely syntactic proofs of properties of elementary axiom systems.

(c) **Complexity of derivations.** We have shown that systems with ground-sequents are equivalent to systems with rules. From the point of view of logic programming, the groundsequents are just programming clauses. It is known that proof search for a clause that has more than one formula at left and at right of the arrow, from a system of such clauses, is not in general polynomial.

6.3 Predicate logic with equality

We shall treat predicate logic with equality as a first example of the extension of sequent calculi by mathematical rules. The main aim is to show that if $\Gamma \to \Delta$ is derivable and contains no equality, the rules for equality are not needed. In other words, predicate logic with equality is conservative over predicate logic without equality.

Axiomatic presentations of predicate logic with equality assume a primitive relation $a = b$ with the axiom of **reflexivity**, $a = a$, and the **replacement scheme**, $a = b \,\&\, A(a/x) \supset A(b/x)$. We showed in Section 2.5 how natural deduction is extended by corresponding rules. Here we shall do the same within sequent calculus.

In the standard treatment in sequent calculus (as in Troelstra and Schwichtenberg 2000, p. 128), one permits derivations to start with sequents of the following form:

Table 6.12 Replacement
through axiomatic sequents

$$\to a = a \qquad a = b, P(a) \to P(b)$$

Here P is an atomic formula. By Gentzen's 'extended Hauptsatz', cuts can be reduced to cuts on axiomatic sequents, but full cut elimination fails. For example, there is no cut-free derivation of symmetry. Weakening and contraction must be assumed.

By our method, cuts on equality axioms are avoided. We first restrict the replacement scheme to atomic predicates $P, Q, R, \ldots$ and then convert the axioms into rules:

Table 6.13 Replacement rules in sequent calculus

$$\frac{a = a, \Gamma \to \Delta}{\Gamma \to \Delta} \; Ref \qquad\qquad \frac{a = b, P(a), P(b), \Gamma \to \Delta}{a = b, P(a), \Gamma \to \Delta} \; Repl$$

When these rules are added to **G3im** and **G3c**, intuitionistic and classical predicate logic with equality are obtained, respectively.

By the restriction to atomic predicates, both forms of rules follow the rule scheme. A case of duplication is produced in the conclusion of the replacement rule if $P(x)$ is $x = b$. The rule where both duplications are contracted is an instance of the reflexivity rule, so the closure condition is satisfied. We therefore have, both for **G3im** and **G3c**, the following theorem:

Theorem 6.14. *The rules of weakening, contraction, and cut are admissible in predicate logic with equality.*

Lemma 6.15. *The replacement axiom $a = b, A(a/x) \to A(b/x)$ is derivable for arbitrary A.*

Proof. By induction on the length of A, using the left and right rules that correspond to the outermost connective of A.   QED.

Lemma 6.16. *The replacement rule*

$$\frac{a = b, A(a/x), A(b/x), \Gamma \to \Delta}{a = b, A(a/x), \Gamma \to \Delta} \ Repl$$

is admissible for an arbitrary predicate A.

Proof. The proof is a sequent calculus version of the proof given in Lemma 2.17.   QED.

Our cut- and contraction-free calculus is equivalent to standard sequent calculi, but the formulation of equality axioms as rules permits proofs by induction on the height of derivation. The conservativity of predicate logic with equality over predicate logic illustrates such proofs. To prove the conservativity, we show that *Ref* can be eliminated from derivations of equality-free sequents.

As observed above, the rule of replacement has an instance with a duplication, and the closure condition is satisfied because the instance in which both duplications are contracted is an instance of reflexivity. For the proof of conservativity, in the absence of *Ref*, the closure condition is satisfied by the addition of the contracted instance of *Repl*:

$$\frac{a = b, b = b, \Gamma \to \Delta}{a = b, \Gamma \to \Delta} \ Repl^*$$

We have the immediate result:

Lemma 6.17. *If $\Gamma \to \Delta$ has no equalities and is derivable in $\mathbf{G3c}$+Ref+ Repl+Repl*, no sequents in its derivation have equalities in the succedent.*

The following lemma contains the essential analysis in the proof of conservativity:

Lemma 6.18. *If $\Gamma \to \Delta$ contains no equalities and is derivable in $\mathbf{G3c}$+Ref+ Repl+Repl*, it is derivable in $\mathbf{G3c}$+Repl+Repl*.*

Proof. We show that all instances of *Ref* can be eliminated from a given derivation, by induction on the height of derivation of a topmost instance

$$\frac{a = a, \Gamma' \to \Delta'}{\Gamma' \to \Delta'} \ Ref$$

If the premiss is an initial sequent also the conclusion is, for by the above lemma, Δ' contains no equality, and the same if it is a conclusion of $L\perp$. If the premiss has been concluded by a logical rule, apply the inductive hypothesis to the premisses and then the rule.

If the premiss has been concluded by *Repl*, there are two cases, according to whether $a = a$ is or is not principal. In the latter case the derivation is, with $\Gamma' = P(b), \Gamma''$,

$$\frac{\dfrac{a = a, b = c, P(b), P(c), \Gamma'' \to \Delta'}{a = a, b = c, P(b), \Gamma'' \to \Delta'} \, \text{\textit{Repl}}}{b = c, P(b), \Gamma'' \to \Delta'} \, \text{\textit{Ref}}$$

By permuting the two rules, the inductive hypothesis can be applied.

If $a = a$ is principal, the derivation is, with $\Gamma' = P(a), \Gamma''$,

$$\frac{\dfrac{a = a, P(a), P(a), \Gamma'' \to \Delta'}{a = a, P(a), \Gamma'' \to \Delta'} \, \text{\textit{Repl}}}{P(a), \Gamma'' \to \Delta'} \, \text{\textit{Ref}}$$

By height-preserving contraction, there is a derivation of $a = a, P(a), \Gamma'' \to \Delta'$ to which the inductive hypothesis applies, with a derivation of $\Gamma' \to \Delta'$ without rule *Ref* as a result.

If the premiss of *Ref* has been concluded by *Repl** with $a = a$ not principal, the derivation is

$$\frac{\dfrac{a = a, b = c, c = c, \Gamma' \to \Delta'}{a = a, b = c, \Gamma' \to \Delta'} \, \text{\textit{Repl*}}}{b = c, \Gamma'' \to \Delta'} \, \text{\textit{Ref}}$$

The rules are permuted and the inductive hypothesis applied.

If $a = a$ is principal, the derivation is

$$\frac{\dfrac{a = a, a = a, \Gamma' \to \Delta'}{a = a, \Gamma' \to \Delta'} \, \text{\textit{Repl*}}}{\Gamma' \to \Delta'} \, \text{\textit{Ref}}$$

Apply now height-preserving contraction and the inductive hypothesis. QED.

Next, because rules *Repl* and *Repl** have equalities in their conclusions, we obtain:

Theorem 6.19. *If $\Gamma \to \Delta$ is derivable in* **G3c**+Ref+Repl+Repl* *and if Γ, Δ contains no equality, then $\Gamma \to \Delta$ is derivable in* **G3c**.

6.4 Herbrand's theorem for universal theories

Let **T** be a theory with a finite number of purely universal axioms and classical logic. We turn the theory **T** into a system of mathematical rules by first removing the quantifiers from each axiom, then converting the remaining part into mathematical rules. The resulting system will be denoted **G3cT**.

Theorem 6.20. Herbrand's theorem. *If the sequent* $\rightarrow \forall x \exists y_1 \ldots \exists y_k A$, *with A quantifier free, is derivable in* **G3cT**, *then there are terms* t_{i_j} *with* $i \leqslant n, j \leqslant k$ *such that*

$$\rightarrow \bigvee_{i=1}^{n} A(t_{i_1}/y_1, \ldots, t_{i_k}/y_k)$$

is derivable in **G3cT**.

Proof. Suppose that $k = 1$. The derivation of $\rightarrow \forall x \exists y A$ ends with

$$\frac{\dfrac{\rightarrow A(z/x, t_1/y), \exists y A(z/x)}{\rightarrow \exists y A(z/x)} \, R\exists}{\rightarrow \forall x \exists y A} \, R\forall$$

Every sequent in the derivation is of the form

$$\Gamma \rightarrow \Delta, A(z/x, t_m/y), \ldots, A(z/x, t_{m+1}/y), \exists y A(z/x)$$

Here Γ, Δ consist of subformulas of $A(z/x, t_i/y)$, with $i < m$, and atomic formulas.

Consider the topsequents of the derivation. If they are initial sequents or conclusions of $L\perp$, they remain so after deletion of the formula $\exists y A(z/x)$. If they are conclusions of zero-premiss mathematical rules, they remain so after the deletion because the right context in these rules is arbitrary. After deletion, every topsequent in the derivation is of the form

$$\Gamma \rightarrow \Delta, A(z/x, t_m/y), \ldots, A(z/x, t_{m+1}/y)$$

The application of the propositional and mathematical inferences as before, but without the formula $\exists y A(z/x)$ in the succedent, produces a derivation of

$$\rightarrow A(z/x, t_1/y), \ldots, A(z/x, t_{m-1}/y), A(z/x, t_m/y), \ldots, A(z/x, t_n/y)$$

and the conclusion follows by applications of rule $R\vee$.   QED.

If the theory **T** is empty we have

Corollary 6.21. *If* $\rightarrow \exists x A$ *is derivable in* **G3c**, *there are terms* $t_1, \ldots, t_n$ *such that* $\rightarrow A(t_1/x) \vee \ldots \vee A(t_n/x)$ *is derivable.*

Notes to Chapter 6

Section 6.1: The use of the word 'sequent' as a noun was begun by Kleene. His *Introduction to Metamathematics* of 1952 (p. 441) explains the origin of the term as follows: 'Gentzen says "Sequenz", which we translate as "sequent", because we have already used "sequence" for any succession of objects, where the German is "Folge".' This is the standard terminology now; Kleene's usage has even been adapted to some other languages with somewhat peculiar results in cases. But Mostowski (1965) for example uses the literal translation 'sequence'.

The classical propositional part of the calculus **G3** was invented by the Finnish logician Oiva Ketonen some time around 1940. It appears first in a publication of 1943 in the Finnish language. Judging from this paper, it seems that Ketonen found his rules as a solution to the problem that anyone who attempts root-first proof search in sequent calculus faces: how does one divide the contexts in two-premiss rules? The obvious answer is that if some assumptions Γ are permitted in the end, they must be permissible elsewhere, and the same for the cases. Thus, there is no need to divide contexts. Other modifications of Gentzen's rules in Ketonen (1944), however, are not as simply explained: he used a single left conjunction and right disjunction rule, where Gentzen had two rules for both, and a left implication rule with shared contexts.

The **G3**-calculi were developed, on the basis of Ketonen's work, by Kleene in his influential book *Introduction to Metamathematics*. A final form of these calculi was given by Dragalin (1988), except for a single succedent intuitionistic version that was found by Troelstra in *Basic Proof Theory*.

The **G0**-calculi, i.e., calculi with independent contexts throughout, were introduced in von Plato (2001b) to give a direct proof of Gentzen's cut-elimination theorem, in place of the original that used a cut rule that wiped out several occurrences of the cut formula in one stroke.

The idea of deducing (in Newton's sense) the rules of sequent calculus from those of natural deduction with general elimination rules is from the first chapter of *Structural Proof Theory*. The interpretation of the structural rules of weakening and contraction in terms of natural deduction was deduced in von Plato (2001a).

Section 6.2: The extension of the **G3**-calculi through mathematical rules, and the proofs of admissibility of the structural rules, was first presented in Negri (1999) for some intuitionistic theories. Extension by rules that correspond to universal axioms was given in Negri and von Plato (1998) which appeared earlier but was written after Negri (1999).

Section 6.3: The treatment of predicate logic with equality, and especially the proof of conservativity of predicate logic with equality over predicate logic, is a result of ours presented in section 6.5 of *Structural Proof Theory*. Anne Troelstra liked this result, shown to him as a manuscript, so much that he asked to include it in the second edition of *Basic Proof Theory* where it now appears as section 4.7.

A different system of sequent calculus for predicate logic with equality is given in Degtyarev and Voronkov (2001). The calculus is an extension of a **G3**-system for classical logic with primitive negation and has the following rules for equality:

$$\frac{}{\Gamma \to \Delta, a = a} \qquad \frac{\Gamma(b/x), a = b \to \Delta(b/x)}{\Gamma(a/x), a = b \to \Delta(a/x)}$$

Similar rules were presented in Wang (1960) (with replacement only in Δ) and in Kanger (1963).

Section 6.4: Our generalization of Herbrand's theorem was inspired by Buss (1995). It appeared first in Negri and von Plato (2005).

7 | Linear order

The extensions of sequent calculi by rules, presented in the previous chapter, share the good structural properties of the purely logical **G3**-calculi, i.e., the rules of weakening, contraction, and cut are admissible. In addition to being admissible, weakening and contraction are height-preserving admissible. The usual consequence of cut elimination, the subformula property, holds in a weaker form, because all the formulas in the derivations in such extensions are subformulas of the endsequent or atomic formulas. However, by analysing, analogously to natural deduction, minimal derivations in specific theories, we can establish a subterm property, by which all terms in a derivation can be restricted to terms in the endsequent.

This chapter gives proofs of the subterm property for partial and linear order, the latter not an easy result. To make its presentation manageable, a system of rules that act on the right part of multisuccedent sequents is used. Further, it is shown through a proof-theoretical algorithm how to linearize a partial order, a result known as Szpilrajn's theorem. The extension is based on the conservativity of the rule system for linear order over that for partial order for sequents that have just one atom in the succedent. Finally, the proof-theoretical solution of the word problem for lattices of Chapter 4 is extended to linear lattices, i.e., lattices in which the order relation is linear.

7.1 Partial order and Szpilrajn's theorem

(a) **Minimal derivations.** We observe that a derivation in which a rule, read root first, produces a duplication of an atom, can be shortened by the application of height-preserving admissibility of contraction in place of the rule that introduces that atom. The possibility of such shortening justifies the following definition:

Definition 7.1. Minimal derivations. *A **minimal** derivation is a derivation in which shortenings through height-preserving admissibility of contraction are not possible, and sequents that can be concluded by zero-premiss rules appear only as topsequents.*

The subterm property, together with the height-preserving admissibility of contraction, will give a bound on proof search for the theories under examination: in a minimal derivation in these theories, no new term can appear, nor any instances of rules that produce a duplication of formulas.

We recall the notion of a Harrop formula in Section 2.3: these formulas are ones that do not contain a disjunction in any positive part. A **Harrop theory** is a theory the axioms of which consist of Harrop formulas. A **left Harrop system** is a system of mathematical rules obtained from the axioms of a Harrop theory by the use of the left rule scheme.

The rules of a left Harrop system have at most one premiss; thus the derivations are linear, not proper trees, and therefore we have:

Theorem 7.2. *If a sequent $\Gamma \rightarrow \Delta$ with only atomic formulas is derivable in a left Harrop system, then $\Gamma \rightarrow P$ is derivable for some atom P. If Δ is not empty, the atom P can be chosen from Δ.*

Proof. Consider a derivation of $\Gamma \rightarrow \Delta$. If the topsequent is an initial sequent $P, \Gamma' \rightarrow \Delta', P$, with $\Delta = \Delta', P$, the succedent can be changed into P. If the topsequent is a zero-premiss mathematical rule, any atom P can be put as the succedent and the derivation with the new succedent continued as with Δ.   QED.

The theorem is similar in nature to Theorem 5.13 by which Skolem's claim about the decidability of universal formulas in lattice theory could be justified. All axioms of lattice theory are Harrop formulas.

(b) Partial order. We consider the theory of partial order as an example of the application of Theorem 7.2. The axioms of partial order are

PO1. $a \leqslant a$,
PO2. $a \leqslant b \mathbin{\&} b \leqslant c \supset a \leqslant c$.

Equality is defined by $a = b \equiv a \leqslant b \mathbin{\&} b \leqslant a$. (Thus, we are working with what are sometimes called preorders or quasiorders.) Clearly, the equality so defined is an equivalence relation and satisfies the principle of substitution of equals in the order relation.

We define **GPO** as a sequent system with the additional rules

Table 7.1 Sequent calculus rules for partial order

$$\frac{a \leqslant a, \Gamma \rightarrow \Delta}{\Gamma \rightarrow \Delta} \ Ref \qquad \frac{a \leqslant c, a \leqslant b, b \leqslant c, \Gamma \rightarrow \Delta}{a \leqslant b, b \leqslant c, \Gamma \rightarrow \Delta} \ Tr$$

The closure condition arises when $a \equiv b$ and $b \equiv c$, so the contracted premiss of rule *Tr* to consider is

$$a \leqslant a, a \leqslant a, \Gamma \rightarrow \Delta$$

The contracted conclusion follows by rule *Ref*, so the closure condition is satisfied without any additions. Thus, the rules of weakening, contraction, and cut are admissible in **GPO**. Weakening and contraction are moreover height-preserving admissible.

Proof analysis in GPO. There are exactly two kinds of derivations to consider. To see what they are, assume that derivations are minimal. If $\Gamma \rightarrow \Delta$ is derivable, the topsequent has the form $P, \Gamma' \rightarrow \Delta', P$ with $\Delta', P = \Delta$, and we can delete Δ'. The two kinds of derivations are:

1. Reflexivity derivations. $P \equiv a \leqslant a$.
The conclusion $\Gamma \rightarrow a \leqslant a$ follows in one step from the initial sequent $a \leqslant a, \Gamma \rightarrow a \leqslant a$, with an application of rule *Ref*:

$$\frac{a \leqslant a, \Gamma \rightarrow a \leqslant a}{\Gamma \rightarrow a \leqslant a} Ref$$

The context Γ is superfluous and can be deleted; thus, the conclusion becomes $\rightarrow a \leqslant a$.

2. Transitivity derivations. The topsequent is $a_1 \leqslant a_n, \Gamma' \rightarrow a_1 \leqslant a_n$.
The atom $a_1 \leqslant a_n$ must be the **removed atom** in a first step of transitivity or else the derivation can be shortened: if some other atom P were removed, with $\Gamma' \equiv P, \Gamma''$, the derivation could be shortened by starting with $a_1 \leqslant a_n, \Gamma'' \rightarrow a_1 \leqslant a_n$ as topsequent.

There cannot be steps of reflexivity in this derivation: the reflexivity atom would be principal in a step of transitivity, or else it could be removed without further ado from the derivation with a subsequent shortening, thus there would be a step of the form

$$\frac{a \leqslant b, a \leqslant a, a \leqslant b, \Gamma \rightarrow a_1 \leqslant a_n}{a \leqslant a, a \leqslant b, \Gamma \rightarrow a_1 \leqslant a_n} Tr$$

By the height-preserving admissibility of contraction, the conclusion of this step could be obtained already from the premiss without using transitivity, in one step less.

Two atoms $a_1 \leqslant a_2$, $a_2 \leqslant a_n$ are **activated** by the step of Tr that removes $a_1 \leqslant a_n$ so that the topsequent is of the form

$$a_1 \leqslant a_n, a_1 \leqslant a_2, a_2 \leqslant a_n, \Gamma'' \to a_1 \leqslant a_n$$

If an atom different from $a_1 \leqslant a_2$ and $a_2 \leqslant a_n$ is removed in the second step, that atom can be deleted from the topsequent. There is then one step less. Therefore, in the second step, one of the activated atoms must become the removed atom, with two new activated atoms, say $a_2 \leqslant a_3$, $a_3 \leqslant a_n$, or else the derivation can be again shortened. The closure of the principal atom $a_1 \leqslant a_n$ with respect to the activation relation gives us what we call a **chain** $a_1 \leqslant a_2, a_2 \leqslant a_3, \ldots a_{n-1} \leqslant a_n$ in the topsequent. By the deletion of the atoms that have not been active in the derivation, we have a derivation of the form

$$\frac{\dfrac{\Gamma''', a_1 \leqslant a_2, a_2 \leqslant a_3, \ldots a_{n-1} \leqslant a_n \to a_1 \leqslant a_n}{\vdots}}{a_1 \leqslant a_2, a_2 \leqslant a_3, \ldots a_{n-1} \leqslant a_n \to a_1 \leqslant a_n} \, Tr$$

in which Γ''' consists of the removed atoms $a_1 \leqslant a_n, \ldots$; thus we have:

Proposition 7.3. *Sequents* $\Gamma \to \Delta$ *derivable in* **GPO** *are derivable as left and right weakenings of reflexivity and transitivity derivations.*

Proof search for a sequent $\Gamma \to \Delta$ is effected by one of the two controls:

Does Δ contain a reflexivity atom?

Does Γ contain a chain from a_1 to a_n with the atom $a_1 \leqslant a_n$ in Δ?

If one of these is the case, the sequent $\Gamma \to \Delta$ is derivable, otherwise it is underivable.

It is seen that a sequent formulation of partial order is harder to analyse than a natural deduction formulation, in which latter case most of the above observations are almost immediate.

Non-degenerate partial order is obtained by adding the axiom

    PO3. $\neg 1 \leqslant 0$.

to PO1 and PO2. The corresponding rule has zero premisses

$$\frac{}{1 \leqslant 0, \Gamma \to \Delta} \, Ndeg$$

Derivations remain linear and the theorem on Harrop systems applies.

If the topsequent is an instance of *Ndeg*, the atom $1 \leqslant 0$ is removed in the next step by *Tr* (it cannot be removed by *Ref*). Steps of *Tr* hide the inconsistent assumption $1 \leqslant 0$, with the general form of conclusion

$$1 \leqslant a_1, a_1 \leqslant a_2, \ldots, a_{n-1} \leqslant 0 \rightarrow a \leqslant b$$

The chain in the antecedent is the closure of formulas activated by $1 \leqslant 0$, and $a \leqslant b$ in the succedent is an arbitrary atom.

Non-trivial partial order has in addition,

> PO4. $0 \leqslant 1$.

The corresponding rule is

$$\frac{0 \leqslant 1, \Gamma \rightarrow \Delta}{\Gamma \rightarrow \Delta} \, Ntriv$$

This rule commutes down with instances of *Tr*. The only interesting case is a transitivity derivation with a chain from which the atom $0 \leqslant 1$ has been removed by *Ntriv*.

(c) Linear order. The theory of linear order is obtained by adding to partial order the **linearity axiom**

> LO. $a \leqslant b \vee b \leqslant a$.

The corresponding rule is

$$\frac{a \leqslant b, \Gamma \rightarrow \Delta \quad b \leqslant a, \Gamma \rightarrow \Delta}{\Gamma \rightarrow \Delta} \, Lin$$

The system of rules for linear order is designated **GLO**.

Theorem 7.4. Conservativity. *If $\Gamma \rightarrow P$ is derivable in* **GLO**, *it is derivable already in* **GPO**.

Proof. Consider a derivation with just one instance of *Lin*, as the last rule, and assume the derivation to be minimal. Thus, the premisses of *Lin* $c \leqslant d, \Gamma \rightarrow P$ and $d \leqslant c, \Gamma \rightarrow P$ are derivable in partial order. If P is a reflexivity atom, $\Gamma \rightarrow P$ is derivable in one step of *Ref*. Otherwise, with $P \equiv a \leqslant b$, there will be two transitive closures of the removed atom $a \leqslant b$ in both derivations of the two premisses of *Lin*, and let them be $a \leqslant a_1, \ldots a_{m-1} \leqslant b$ and $a \leqslant b_1, \ldots b_{n-1} \leqslant b$. If $c \leqslant d$ is not an atom in the first chain, it can be deleted and a derivation of $\Gamma \rightarrow P$ in partial order obtained, and similarly for $d \leqslant c$ in the second chain. Thus, we have the two chains

$$a \leqslant a_1, \ldots, a_i \leqslant c, c \leqslant d, d \leqslant a_{i+1}, \ldots a_{m-1} \leqslant b$$

$$a \leqslant b_1, \ldots, b_j \leqslant d, d \leqslant c, c \leqslant b_{j+1}, \ldots b_{n-1} \leqslant b$$

A chain from a to b can be constructed in two ways, say,

$$a \leqslant a_1, \ldots, a_i \leqslant c, c \leqslant b_{j+1}, \ldots b_{n-1} \leqslant b$$

Now the sequent

$$a \leqslant a_1, \ldots, a_i \leqslant c, c \leqslant b_{j+1}, \ldots b_{n-1} \leqslant b \rightarrow a \leqslant b$$

is derivable in partial order.   QED.

The conservativity theorem extends to non-degenerate non-trivial partial order (note that non-triviality follows from non-degeneracy if the order is linear: $\neg\, 1 \leqslant 0$ and $0 \leqslant 1 \vee 1 \leqslant 0$ give $0 \leqslant 1$):

Theorem 7.5. *If $\Gamma \rightarrow P$ is derivable in non-degenerate* **GLO***, it is already derivable in non-degenerate non-trivial* **GPO***.*

(d) Extension algorithm from partial to linear order.

Definition 7.6. *An ordering Σ is* **inconsistent** *if $\Gamma \rightarrow 1 \leqslant 0$ is derivable for some finite subset Γ of Σ, otherwise it is* **consistent***.*

Corollary 7.7. Szpilrajn's theorem. *Given a set Σ of atoms in a consistent non-degenerate partial ordering, it can be extended to a consistent non-degenerate linear ordering.*

Proof. Let a, b be any two elements in Σ not ordered in Σ. We claim that either $\Sigma, a \leqslant b$ or $\Sigma, b \leqslant a$ is consistent in **GPO**. Let us assume the contrary, i.e., that there exists a finite subset Γ of Σ such that both $\Gamma, a \leqslant b \rightarrow 1 \leqslant 0$ and $\Gamma, b \leqslant a \rightarrow 1 \leqslant 0$ are derivable in **GPO**. We then have the step

$$\frac{a \leqslant b, \Gamma \rightarrow 1 \leqslant 0 \quad b \leqslant a, \Gamma \rightarrow 1 \leqslant 0}{\Gamma \rightarrow 1 \leqslant 0} \, Lin$$

Now $\Gamma \rightarrow 1 \leqslant 0$ is derivable in **GLO**, and by the conservativity theorem, $\Gamma \rightarrow 1 \leqslant 0$ is already derivable in **GPO**, contrary to the consistency assumption. Iteration of the procedure gives the desired extension.   QED.

Remark. Constructive conservativity vs non-constructive extension. The proof of the above conservativity theorem is constructive, and effectivity of the extension depends on how the set Σ is given.

We observe a more general phenomenon: classical set-theoretic extension results that use non-constructive principles such as Zorn's lemma are reformulated as constructive proof-theoretical conservativity results. An example is the constructive conservativity of linear order over partial order vs Szpilrajn's (1930) classical theorem. Another example is the pointfree constructive Hahn-Banach theorem (Cederquist, Coquand, and Negri 1998) vs the classical, non-constructive Hahn-Banach theorem.

Decidability of the order relation is often assumed, either explicitly or through the application of the law of excluded middle. Our treatment does not impose any such requirement and therefore does not rule out a computational approach to order relations in continuous sets.

The law of excluded middle is avoided by considering extensions of the intuitionistic calculus **G3im**, instead of the classical one.

Observe that the intuitionistic rules of implication do not permute down with mathematical rules if these latter have at least two premises. In the case of Harrop theories, such as partial order or lattice theory, logical rules do permute down and derivations by mathematical rules can be considered in isolation. The separation of the logical and mathematical parts of derivations holds with no restrictions if classical propositional logic is used, because of the invertibility of all its rules.

In the previous chapter, we developed proof analyses for the **G3** sequent calculus and for the mathematical rules formulated in the form of a left rule scheme. The question arises whether we can change the basic calculus, or the form of the rule scheme, or both. The answer is positive, but some care is needed to guarantee the admissibility of the structural rules in the extended calculi. In general, the form of the rule scheme will have to be in harmony with the basic calculus. If, for instance, we should modify the basic calculus in favour of context-independent rules, the rule scheme would have to be context-independent as well.

7.2 The word problem for linear order

In this section we shall introduce a variant of proof analysis with an application to a specific problem, the word problem for linear order. We shall first present a dual of the left rule scheme, namely the right rule scheme, and show how it works in the example of the theory of linear order. For Harrop theories, the right rule scheme has single succedent sequents and can be presented as an extension of natural deduction, as in Chapter 2. The example of lattice theory shows how the method of permutation of rules works for systems of this form. In order to extend the method of permutation of rules to non-Harrop theories, such as the theory of linear lattices, one would need a multiple-conclusion system of natural deduction. We shall see in Chapter 10 how such a calculus works. One aspect is that derivations by multiple-conclusion rules cannot always be written in the form of two-dimensional trees, but the difficulty will be circumvented by the use of sequent systems.

(a) **Systems of right rules.** The **left** rule scheme for an axiom of the form

$$P_1 \& \ldots \& P_m \supset Q_1 \lor \ldots \lor Q_n$$

is as follows:

$$\frac{Q_1, P_1, \ldots, P_m, \Gamma \to \Delta \quad \ldots \quad Q_n, P_1, \ldots, P_m, \Gamma \to \Delta}{P_1, \ldots, P_m, \Gamma \to \Delta}$$

It has a dual formulation as a **right** rule scheme:

Table 7.2 The right rule scheme

$$\frac{\Gamma \to \Delta, Q_1, \ldots, Q_n, P_1 \quad \ldots \quad \Gamma \to \Delta, Q_1, \ldots, Q_n, P_m}{\Gamma \to \Delta, Q_1, \ldots, Q_n}$$

We have a repetition of the atoms Q_i in the premisses, to obtain admissibility of right contraction.

As for the left rule scheme, we have the following condition:

Closure condition. *If the atoms in a rule have an instance that makes two atoms in the conclusion identical, the contracted rule has to be added.*

The analogue of Theorem 6.9 holds for extensions based on the right rule scheme:

Theorem 7.8. *The structural rules of left and right weakening and contraction are height-preserving and the rule of cut admissible in extensions of* **G3c** *and* **G3im** *by rules that follow the right rule scheme and satisfy the closure condition.*

(b) **Linear order.** We consider the theory of linear order as a system with right rules.

Table 7.3 Right rules for linear order

$$\frac{}{\Gamma \rightarrow \Delta, a \leqslant b, b \leqslant a} \, Lin \qquad \frac{}{\Gamma \rightarrow \Delta, a \leqslant a} \, Ref$$

$$\frac{\Gamma \rightarrow \Delta, a \leqslant c, a \leqslant b, \quad \Gamma \rightarrow \Delta, a \leqslant c, b \leqslant c}{\Gamma \rightarrow \Delta, a \leqslant c} \, Tr$$

The term b in rule Tr is called a **middle term**.

The following result establishes the subterm property for the theory of linear order.

Theorem 7.9. *All terms in a minimal derivation of $\Gamma \rightarrow \Delta$ in the right theory of linear order are terms in Γ, Δ.*

Proof. We show first that rule Ref need not be considered: if a topsequent is an instance of Ref, the first step must be a step of Tr that removes a reflexivity atom $a \leqslant a$. The derivation has the form

$$\frac{\Gamma \rightarrow \Delta, a \leqslant c, a \leqslant a, \quad \Gamma \rightarrow \Delta, a \leqslant c, a \leqslant c}{\Gamma \rightarrow \Delta, a \leqslant c} \, Tr$$

The conclusion follows from the right premiss by height-preserving contraction, contrary to the assumption of a shortest derivation. Thus, proper derivations start with initial sequents or instances of Lin, followed by instances of Tr.

Let b be a first middle term from top that disappears from the derivation in a step of transitivity, and we may assume this to be the last step. We show that the derivation can be shortened. We have the instance

$$\frac{\Gamma \overset{\vdots}{\rightarrow} \Delta, a \leqslant c, a \leqslant b, \quad \Gamma \overset{\vdots}{\rightarrow} \Delta, a \leqslant c, b \leqslant c}{\Gamma \rightarrow \Delta, a \leqslant c} \, Tr$$

If $a \leqslant b$ is never active in the rightmost branch of the derivation that leads to the left premiss, it can be deleted and the derivation shortened. Tracing up along the right branch from $a \leqslant b$, we find a removed atom $d \leqslant b$, and we continue tracing the atoms removed in steps of transitivity that have the previously traced atom as principal, until we arrive at an atom $e \leqslant b$ in a topsequent. It is not principal in an initial sequent because the term b would appear in the antecedent. If it is not principal in Lin, it is deleted together with the step of Tr that removes it. If it is principal in Lin, the topsequent is of the form

$$\Gamma \rightarrow \Delta', e \leqslant b, b \leqslant e$$

There must be a step that removes $b \leqslant e$. Because a right branch was followed, there is a step with a removed atom $f \leqslant b$ in a left premiss:

$$\frac{\overset{\vdots}{\Gamma \to \Delta'', f \leqslant e, f \leqslant b,} \quad \overset{\vdots}{\Gamma \to \Delta'', f \leqslant e, b \leqslant e}}{\underset{\vdots}{\Gamma \to \Delta'', f \leqslant e}} \; Tr$$

Tracing $f \leqslant b$ up the rightmost branch in the same way as $a \leqslant b$, we find a topsequent with an atom $g \leqslant b$. Now an argument as for the atom $e \leqslant b$ applies, and for the process to stop at some finite stage, we must find an atom with b not principal in *Lin*. It can be deleted and the derivation shortened.   QED.

The decidability of universal formulas reduces to the derivability of a finite number of implications of the form

$$P_1 \& \ldots \& P_m \supset Q_1 \vee \ldots \vee Q_n.$$

These are derivable in a system of rules if and only if the sequent

$$P_1, \ldots, P_m \to Q_1, \ldots, Q_n$$

is derivable. By Theorem 7.9, proof search for such a sequent in the theory of linear order can be restricted to atoms with known terms and to minimal derivations, which makes proof search bounded and decidable. We therefore have the following corollary:

Corollary 7.10. *The quantifier-free theory of linear order is decidable.*

Proof. Application of rule *Tr*, root first, with middle terms chosen from the conclusion, can produce only a bounded number of distinct atoms in the premisses. Whenever a duplication is produced, proof search fails by the height-preserving admissibility of contraction.   QED.

The first-order theory of linear order has been shown decidable in earlier literature, but only in a weak sense: the theorems are recursively enumerable because the theory is formalized, but it has been proved that the non-theorems are also recursively enumerable. Such a result gives a decision algorithm, but with no upper bound, whereas the above gives a decision algorithm for universal formulas with a bound on proof search.

7.3 Linear lattices

The theory of linear lattices has a binary partial order relation $a \leqslant b$ and equality is defined by $a = b \equiv a \leqslant b \,\&\, b \leqslant a$.

The axioms are as follows:

Table 7.4 The axioms for a linear lattice

$a \leqslant a$	(Ref),	$a \leqslant b \vee b \leqslant a$	(Lin),	$a \leqslant b \,\&\, b \leqslant c \supset a \leqslant c$	(Tr),
$a \wedge b \leqslant a$	(L$\wedge_1$),	$a \wedge b \leqslant b$	(L$\wedge_2$),	$c \leqslant a \,\&\, c \leqslant b \supset c \leqslant a \wedge b$	(R$\wedge$),
$a \leqslant a \vee b$	(R$\vee_1$),	$b \leqslant a \vee b$	(R$\vee_2$),	$a \leqslant c \,\&\, b \leqslant c \supset a \vee b \leqslant c$	(L$\vee$).

The principle of substitution of equals in the lattice operations can be proved, because equality is defined through the partial order relation.

We start by the observation that in linear lattices, in addition to the lattice equivalences, $a \vee b \leqslant c$ if and only if $a \leqslant c$ and $b \leqslant c$ and $c \leqslant a \wedge b$ if and only if $c \leqslant a$ and $c \leqslant b$, we also have the equivalences that hold in virtue of linearity, $a \wedge b \leqslant c$ if and only if $a \leqslant c$ or $b \leqslant c$ and $c \leqslant a \vee b$ if and only if $c \leqslant a$ or $c \leqslant b$. These equivalences will lead to a sequent calculus proof system for linear lattices. In establishing the structural properties of the calculus, we shall also use the fact that reflexivity and linearity can be restricted to **ground terms**, defined as follows:

Definition 7.11. *A* **ground term** *is one that does not contain lattice operations.*

Ground terms will be denoted by $p, q, r, \ldots$ The rules of our calculus for linear lattices are the following:

Table 7.5 A rule system for linear lattices

$$\frac{}{\Gamma \to \Delta, p \leqslant p} \; Ref \qquad \frac{\Gamma \to \Delta, p \leqslant q, q \leqslant p}{} \; Lin$$

$$\frac{\Gamma \to \Delta, a \leqslant c, a \leqslant b \quad \Gamma \to \Delta, a \leqslant c, b \leqslant c}{\Gamma \to \Delta, a \leqslant c} \; Tr$$

$$\frac{c \leqslant a, c \leqslant b, \Gamma \to \Delta}{c \leqslant a \wedge b, \Gamma \to \Delta} \; L{\wedge}R \qquad \frac{a \leqslant c, \Gamma \to \Delta \quad b \leqslant c, \Gamma \to \Delta}{a \wedge b \leqslant c, \Gamma \to \Delta} \; L{\wedge}L$$

$$\frac{\Gamma \to \Delta, c \leqslant a \quad \Gamma \to \Delta, c \leqslant b}{\Gamma \to \Delta, c \leqslant a \wedge b} \; R{\wedge}R \qquad \frac{\Gamma \to \Delta, a \leqslant c, b \leqslant c}{\Gamma \to \Delta, a \wedge b \leqslant c} \; R{\wedge}L$$

$$\frac{\Gamma \to \Delta, c \leqslant a, c \leqslant b}{\Gamma \to \Delta, c \leqslant a \vee b} \; R{\vee}R \qquad \frac{\Gamma \to \Delta, a \leqslant c \quad \Gamma \to \Delta, b \leqslant c}{\Gamma \to \Delta, a \vee b \leqslant c} \; R{\vee}L$$

$$\frac{c \leqslant a, \Gamma \to \Delta \quad c \leqslant b, \Gamma \to \Delta}{c \leqslant a \vee b, \Gamma \to \Delta} \; L{\vee}R \qquad \frac{a \leqslant c, b \leqslant c, \Gamma \to \Delta}{a \vee b \leqslant c, \Gamma \to \Delta} \; L{\wedge}L$$

Derivations start with initial sequents of the form $a \leqslant b, \Gamma \to \Delta, a \leqslant b$ and with instances of the zero-premiss rules. Of these rules, *Ref* and *Lin* are restricted to ground terms.

An inspection of the rules shows that middle terms in *Tr* are the only terms in premisses that need not be also terms in a conclusion. By the permutability of logical rules past the mathematical rules, we consider derivations of sequents with only atomic formulas.

The above rules give a complete system for the theory of linear lattices, because reflexivity and linearity are derivable for arbitrary terms:

Lemma 7.12. *The sequents* $\to a \leqslant a$ *and* $\to a \leqslant b, b \leqslant a$ *are derivable for arbitrary terms a and b in the rule system for linear lattices.*

Proof. By induction on the length of the terms a, b. For ground terms the sequents are zero-premiss rules of the system, thus derivable. For a compound term a, for instance $a \equiv a_1 \wedge a_2$, reflexivity follows from the meet rules. We get by weakening from the inductive hypothesis $\to a_1 \leqslant a_1$, $a_2 \leqslant a_1$ and $\to a_1 \leqslant a_2, a_2 \leqslant a_2$. These give, both by $R \wedge R$, $\to a_1 \wedge a_2 \leqslant a_1$ and $\to a_1 \wedge a_2 \leqslant a_2$, and the conclusion $\to a_1 \wedge a_2 \leqslant a_1 \wedge a_2$ follows by $R \wedge R$.

If a is a join, the proof uses instead the rules for join.

For linearity, we have to analyse the form of a and b. If a and b are not both ground terms, there are eight cases, reduced to five by symmetry. In all such cases, linearity is reduced to linearity on the components that is derivable by the inductive hypothesis. For instance, in the case where $a \equiv a_1 \wedge a_2$, $b \equiv b_1 \vee b_2$, linearity on a, b is derived by applying $R \wedge R$ to the sequents $\to a_1 \wedge a_2 \leqslant b_1 \vee b_2, b_1 \vee b_2 \leqslant a_1$ and $\to a_1 \wedge a_2 \leqslant b_1 \vee b_2, b_1 \vee b_2 \leqslant a_2$. The former is derived by $R \vee L$ from

$$\frac{\dfrac{\to a_1 \leqslant b_1, a_1 \leqslant b_2, a_2 \leqslant b_1 \vee b_2, b_1 \leqslant a_1}{\to a_1 \leqslant b_1 \vee b_2, a_2 \leqslant b_1 \vee b_2, b_1 \leqslant a_1} R \vee R}{\to a_1 \wedge a_2 \leqslant b_1 \vee b_2, b_1 \leqslant a_1} R \wedge L$$

and

$$\frac{\dfrac{\to a_1 \leqslant b_1, a_1 \leqslant b_2, a_2 \leqslant b_1 \vee b_2, b_2 \leqslant a_1}{\to a_1 \leqslant b_1 \vee b_2, a_2 \leqslant b_1 \vee b_2, b_2 \leqslant a_1} R \vee R}{\to a_1 \wedge a_2 \leqslant b_1 \vee b_2, b_2 \leqslant a_1} R \wedge L$$

Here the topsequents are derivable by the inductive hypothesis. The latter is derived in a similar way.   QED.

The following results show that the system of sequent calculus for linear lattices has the same structural properties as the purely logical calculus **G3c**.

Lemma 7.13. *All the rules of the system for linear lattices are height-preserving invertible.*

Proof. By induction on the height of the derivation.   QED.

Proposition 7.14. *Weakening and contraction are height-preserving admissible in the rule system for linear lattices.*

Proof. The proof follows the structure of the proof of admissibility of the structural rules for extensions of **G3c**. Observe that, because of the invertibility of the lattice rules, contraction on atomic formulas with lattice structure gets reduced to contraction on smaller atomic formulas. For this reason, unlike in the rules of the general rule scheme, there is no need to repeat the principal atoms of the rule in its premisses.   QED.

Theorem 7.15. *The rule of cut is admissible in the rule system for linear lattices.*

Proof. The proof follows the usual pattern of the proof of cut elimination for extensions of **G3c**, with a refined measure on atomic formulas that takes into account the complexity of terms in them. By the permutation of the logical rules below the mathematical rules, and the permutation of cut with respect to the latter, we can limit our analysis to the part of the derivation with only mathematical rules. Observe that to consider the system has 11 mathematical rules so there are *a priori* 121 different cases of cut with cut formula principal in both premisses. Of these, there are pairs that get excluded because they are both right rules or both left rules, others that get excluded because the terms do not match, and those with reflexivity and linearity with lattice rules that are excluded because reflexivity and linearity are restricted to ground terms. In the end, we are left with 15 cases that we consider below:

1. The cut formula $a \leqslant b \wedge c$ is principal in both premisses of cut and both are derived by meet rules. We have three subcases according to the rules used to derive the premisses of cut.

   1.1. The derivation is

$$\frac{\dfrac{\Gamma \to \Delta, a \leqslant b \quad \Gamma \to \Delta, a \leqslant c}{\Gamma \to \Delta, a \leqslant b \wedge c} R\wedge R \quad \dfrac{a \leqslant b, a \leqslant c, \Gamma' \to \Delta'}{a \leqslant b \wedge c, \Gamma' \to \Delta'} L\wedge R}{\Gamma, \Gamma' \to \Delta, \Delta'} Cut$$

It is transformed into

$$
\dfrac{\Gamma \to \Delta, a \leqslant c \quad \dfrac{\Gamma \to \Delta, a \leqslant b \quad a \leqslant b, a \leqslant c, \Gamma' \to \Delta'}{a \leqslant c, \Gamma, \Gamma' \to \Delta, \Delta'} \; Cut}{\dfrac{\Gamma, \Gamma, \Gamma' \to \Delta, \Delta, \Delta'}{\Gamma, \Gamma' \to \Delta, \Delta'} \; Ctr^*} \; Cut
$$

Both cuts are on smaller atomic formulas.

1.2. The term a is $a_1 \wedge a_2$ and the derivation is

$$
\dfrac{\dfrac{\Gamma \to \Delta, a \leqslant b \quad \Gamma \to \Delta, a \leqslant c}{\Gamma \to \Delta, a \leqslant b \wedge c} \; R\wedge R \quad \dfrac{a_1 \leqslant b \wedge c, \Gamma' \to \Delta' \quad a_2 \leqslant b \wedge c, \Gamma' \to \Delta'}{a_1 \wedge a_2 \leqslant b \wedge c, \Gamma' \to \Delta'} \; L\wedge L}{\Gamma, \Gamma' \to \Delta, \Delta'} \; Cut
$$

It is transformed as follows

$$
\dfrac{\dfrac{\dfrac{\Gamma \to \Delta, a_1 \wedge a_2 \leqslant b \wedge c}{\Gamma \to \Delta, a_1 \leqslant b \wedge c, a_2 \leqslant b \wedge c} \; R\wedge L\text{-}Inv \quad a_1 \leqslant b \wedge c, \Gamma' \to \Delta'}{\Gamma, \Gamma' \to \Delta, \Delta', a_2 \leqslant b \wedge c} \; Cut \quad a_2 \leqslant b \wedge c, \Gamma' \to \Delta'}{\dfrac{\Gamma, \Gamma', \Gamma' \to \Delta, \Delta', \Delta'}{\Gamma, \Gamma' \to \Delta, \Delta'} \; Ctr^*} \; Cut
$$

Here $R\wedge L$-*Inv* is the (height-preserving admissible) inversion of rule $R\wedge L$ and both cuts are on smaller cut formulas.

1.3. The derivation is

$$
\dfrac{\dfrac{\Gamma \to \Delta, b \leqslant a, c \leqslant a}{\Gamma \to \Delta, b \wedge c \leqslant a} \; R\wedge L \quad \dfrac{b \leqslant a, \Gamma' \to \Delta' \quad c \leqslant a, \Gamma' \to \Delta'}{b \wedge c \leqslant a, \Gamma' \to \Delta'} \; L\wedge L}{\Gamma, \Gamma' \to \Delta, \Delta'} \; Cut
$$

and the transformed derivation is

$$
\dfrac{\dfrac{\dfrac{\Gamma \to \Delta, b \leqslant a, c \leqslant a \quad b \leqslant a, \Gamma' \to \Delta'}{\Gamma, \Gamma' \to \Delta, \Delta', c \leqslant a} \; Cut \quad c \leqslant a, \Gamma' \to \Delta'}{\Gamma, \Gamma', \Gamma' \to \Delta, \Delta', \Delta'} \; Cut}{\Gamma, \Gamma' \to \Delta, \Delta'} \; Ctr^*
$$

with two cuts on smaller formulas.

2. One of the premisses of cut is derived by a meet rule and the other by a join rule. There are four subcases ($R\wedge R/L\vee L$, $R\wedge L/L\vee R$, $R\vee R/L\vee R$, $R\vee L/L\wedge L$). We shall consider in detail only the first, the others being similar.

The derivation is of the form, with $a \equiv a_1 \vee a_2$,

$$
\dfrac{\dfrac{\Gamma \to \Delta, a \leqslant b \quad \Gamma \to \Delta, a \leqslant c}{\Gamma \to \Delta, a \leqslant b \wedge c} \; R\wedge R \quad \dfrac{a_1 \leqslant b \wedge c, a_2 \leqslant b \wedge c, \Gamma' \to \Delta'}{a_1 \vee a_2 \leqslant b \wedge c, \Gamma' \to \Delta'} \; L\vee L}{\Gamma, \Gamma' \to \Delta, \Delta'} \; Cut
$$

It is converted into

$$\dfrac{\dfrac{\dfrac{a_1 \leqslant b \wedge c, a_2 \leqslant b \wedge c, \Gamma' \to \Delta'}{a \leqslant b \wedge c, \Gamma' \to \Delta'} \; L\vee L\text{-}Inv}{\Gamma \to \Delta, a \leqslant b \quad \dfrac{a \leqslant b, a \leqslant c, \Gamma' \to \Delta'}{a \leqslant c, \Gamma, \Gamma' \to \Delta, \Delta'} \; Cut} \; L\wedge R\text{-}Inv}{\Gamma \to \Delta, a \leqslant c \quad \dfrac{\;}{\;}}$$

$$\dfrac{\Gamma, \Gamma, \Gamma' \to \Delta, \Delta, \Delta'}{\Gamma, \Gamma' \to \Delta, \Delta'} \; Ctr^*$$

There is a cut of smaller height followed by a cut on a smaller formula.

3. Both premisses of cut are derived by join rules ($R \vee L/L \vee L$, $R \vee L/L \vee R$, $R \vee R/L \vee L$, $R \vee R/L \vee R$). These cases are all variants of the cases considered above.

4. One of the premisses of cut is *Tr* and the other is a lattice rule. We have four subcases, for the left meet and join rules. We consider the following case, with $c \equiv c_1 \wedge c_2$:

$$\dfrac{\dfrac{\Gamma, \to \Delta, a \leqslant c, a \leqslant b \quad \Gamma, \to \Delta, a \leqslant c, b \leqslant c}{\Gamma, \to \Delta, a \leqslant c} \; Tr \quad \dfrac{a \leqslant c_1, a \leqslant c_2, \Gamma' \to \Delta'}{a \leqslant c_1 \wedge c_2, \Gamma' \to \Delta'} \; L\wedge R}{\Gamma, \Gamma' \to \Delta, \Delta'} \; Cut$$

It is converted into

$$\dfrac{\dfrac{\Gamma, \to \Delta, a \leqslant c}{\Gamma, \to \Delta, a \leqslant c_2} \; R\wedge R\text{-}Inv \quad \dfrac{\dfrac{\Gamma, \to \Delta, a \leqslant c}{\Gamma, \to \Delta, a \leqslant c_1} \; R\wedge R\text{-}Inv \quad a \leqslant c_1, a \leqslant c_2, \Gamma' \to \Delta'}{a \leqslant c_2, \Gamma, \Gamma' \to \Delta, \Delta'} \; Cut}{\dfrac{\Gamma, \Gamma, \Gamma' \to \Delta, \Delta, \Delta'}{\Gamma, \Gamma' \to \Delta, \Delta'} \; Ctr^*} \; Cut$$

There are two cuts on smaller cut formulas. Observe that *Tr* has disappeared. All the remaining cases are obtained in a similar way by the use of inversions and reduction of cuts to cuts on smaller formulas.   QED.

The following result permits us to reduce the word problem for linear lattices to that for linear order:

Proposition 7.16. *Rule* Tr *permutes up with respect to the lattice rules.*

Proof. Consider the topmost transitivity and the rules used to derive its premisses. If one of them is *Ref*, the conclusion is equal to the other premiss and the step can be dispensed with. If one is *Lin* and the other a lattice rule, say, for example, $R \wedge L$, we have a derivation of the form

$$\dfrac{\dfrac{\Gamma \to \Delta, a_1 \leqslant p, a_2 \leqslant p, q \leqslant p}{\Gamma \to \Delta, a_1 \wedge a_2 \leqslant p, q \leqslant p} \; R\wedge L \quad \dfrac{\Gamma \to \Delta, p \leqslant q, q \leqslant p}{\;} \; Lin}{\Gamma \to \Delta, a_1 \wedge a_2 \leqslant q, q \leqslant p} \; Tr$$

First weaken the left topsequent by the atom $a_2 \leqslant q$. To fit the derivation into a page's breadth, we leave unwritten the contexts Γ, Δ and then have the transformed derivation:

$$\cfrac{\cfrac{\to a_1 \leqslant p, a_2 \leqslant q, a_2 \leqslant p, q \leqslant p \quad \overline{\to a_1 \leqslant p, a_2 \leqslant q, p \leqslant q, q \leqslant p}\ \text{Lin}}{\to a_1 \leqslant p, a_2 \leqslant q, q \leqslant p}\ \text{Tr} \quad \cfrac{\overline{\to a_1 \leqslant p, a_2 \leqslant q, p \leqslant q, q \leqslant p}\ \text{Lin}}{}\ \text{Tr}}{\cfrac{\to a_1 \leqslant q, a_2 \leqslant q, q \leqslant p}{\to a_1 \wedge a_2 \leqslant q, q \leqslant p}\ R{\wedge}L}$$

If both premisses of transitivity are obtained by lattice rules, it suffices to consider the derivation of one of the premisses. If the middle term of transitivity gets decomposed by the lattice rule, inversion is applied to the other premiss of transitivity in order to move transitivity up. Otherwise if the middle term is unaffected by the lattice rule, transitivity is simply permuted up with respect to that rule.   QED.

By the permutation result, we can collect our analyses of linear order and linear lattices into the following:

Proposition 7.17. Structure of derivations in linear lattices. *Derivations in the rule system for linear lattices can be so transformed as to have a part of the derivation in the rule system for linear order, followed by rules for the lattice operations, followed by the logical rules.*

It follows, in particular, that rule *Tr* can be restricted to ground terms. We thus obtain the following:

Corollary 7.18. Subterm property. *If a sequent is derivable in the theory of linear lattices, it has a derivation with no new terms.*

Proof. By Proposition 7.16, we can assume that the derivation has a form in which no occurrence of transitivity follows applications of rules for lattice operations. Any new term thus belongs to the part of the derivation in the system for linear order. The result follows by Theorem 7.9.   QED.

Notes to Chapter 7

The proof-theoretical treatment of Szpilrajn's theorem in Section 7.1 comes from Negri, von Plato, and Coquand (2004), as comes the possibility of a right rule scheme and the proof of the subterm property for the theory of linear order.

There is an interesting prehistory to the decision problem of the theory of linear order of Section 7.2. It was announced as an open problem in Janisczak (1953). In a review of this paper, Georg Kreisel (1954) announced a positive solution to the first-order decision problem and made a brief sketch of it. It turned out, however, to be defective, because Kreisel himself published a correction to this effect in the end of the 1954 volume of *Mathematical Reviews*. (We thank Ryan Siders for having spotted the correction, not visible in the online version.) Andrzej Ehrenfeucht announced in (1959) a positive solution, but gave no proof. Such a proof was given by Läuchli and Leonard (1966), however, only in the weak sense of, one could say, impossibility of undecidability, as explained in Section 7.2.

Section 7.3 uses a system of rules for lattice operations that was developed from a suggestion by Alex Simpson, after the first author had presented another multisuccedent system for linear lattices, obtained as an extension of the system for lattices of Section 4.2. The suggestion was to exploit the equivalences $a \wedge b \leqslant c$ if and only if $a \leqslant c$ or $b \leqslant c$ and $c \leqslant a \vee b$ if and only if $c \leqslant a$ or $c \leqslant b$ that hold in a linear lattice. Simpson's view was that the word problem for linear lattices should have a simpler, not a more complicated, solution than that for lattices in general, and indeed, compared to the above proof, the proof of the subterm property in Negri (2005b) is much more complicated. A further development of the methods of Section 7.3 into a positive solution of the uniform word problem for linear Heyting algebras is found in Dyckhoff and Negri (2006). Such algebras are interesting from a logical point of view because they function as models of what is known as Dummett logic; cf. chapter 7 of *Structural Proof Theory*.

The axioms of a Heyting algebra, without the linearity axiom, do not contain disjunctions and can be converted into rules in natural deduction style. It is, however, unlikely that derivations by these rules could have a subterm property. Were this the case, there would be a polynomial bound on proof search in propositional logic.

PART III

Proof systems for geometric theories

8 | Geometric theories

8.1 Systems of geometric rules

(a) **The geometric rule scheme.** We used first natural deduction as the logical calculus that is extended by rules. The axioms covered were formulas without essential disjunctions. With sequent calculus as the logical calculus to be extended, any universal axioms could be converted into rules. In this chapter, we shall show how the class of axioms convertible into rules is further extended into **geometric implications**, as in Chapter 5, but without the restrictions that natural deduction brings with itself. We recall the definition of a geometric implication:

Definition 8.1. *A formula in first-order logic is* **geometric** *if it does not contain* $\supset$ *or* $\forall$. *A* **geometric implication** *has the form, with A and B geometric formulas,*

$$\forall x \ldots \forall z (A \supset B).$$

A **geometric theory** *is a theory axiomatized by geometric implications.*

The examples of geometric axioms mostly encountered in mathematics are **existential axioms**. An existential axiom can replace a construction and postulated properties of constructed objects. It has in this case the form $\forall x \ldots \forall y \exists z A(x, \ldots, y, z)$ and corresponds to the construction of some z from any given $x, \ldots, y$, such that $A(x, \ldots, y, z)$ holds. Constructions can have conditions, as in elementary geometry where an intersection point of two lines, say, can be constructed only if the lines are convergent. The existential axiom together with the condition can be given, when suitably formulated, as a geometric implication.

In lattice theory, the propositional part of existential axioms consists of atomic formulas $P(x, y, z)$. The rule that corresponds to such an existential axiom is, with parameters a, b in place of the universally quantified variables:

$$\frac{P(a, b, z), \Gamma \to \Delta}{\Gamma \to \Delta} \text{ E-Rule}$$

The rule has the variable restriction that the eigenvariable z must not be free in the conclusion. Assuming the premiss of the rule, application of the logical rule $L\exists$ followed by $L\forall$ twice gives as a conclusion

$$\forall x \forall y \exists z P(x, y, z), \Gamma \to \Delta$$

A cut with the axiomatic sequent $\to \forall x \forall y \exists z P(x, y, z)$ gives the conclusion of the rule. On the other hand, application of $R\exists$ to the initial sequent $P(x, y, z) \to P(x, y, z)$ followed by the rule and universal generalizations gives the axiom, so it follows that the rule has the same force as the existential axiom.

We shall give in the next Subsection (b) concrete examples of theories axiomatized by geometric implications.

Proposition 8.2. Canonical form for geometric implications. *Geometric implications can be reduced to conjunctions of formulas of the form*

$$\forall \bar{x}(P_1 \& \ldots \& P_m \supset \exists \bar{y}_1 M_1 \vee \ldots \vee \exists \bar{y}_n M_n)$$

Each P_i is an atomic formula, each M_j is a conjunction of a list of atomic formulas $\overline{Q}_j$, and none of the variables in the vectors $\bar{y}_j$ are free in P_i.

The **geometric rule scheme** that corresponds to geometric axioms has the form

Table 8.1 The geometric rule scheme

$$\frac{\overline{Q}_1(\bar{z}_1/\bar{y}_1), \overline{P}, \Gamma \to \Delta \quad \ldots \quad \overline{Q}_n(\bar{z}_n/\bar{y}_n), \overline{P}, \Gamma \to \Delta}{\overline{P}, \Gamma \to \Delta} \; GRS$$

The variables $\bar{y}_i$ are called the **replaced variables** of the scheme, and the variables $\bar{z}_i$ the **proper variables**, or **eigenvariables**. In what follows, we shall consider for ease of notation the case in which the vectors of variables $\bar{y}_i$ consist of a single variable. All the proofs can be adapted in a straightforward way to the general case.

The geometric rule scheme is subject to the following condition that expresses in a logic-free way the role of the existential quantifier in a geometric axiom:

Condition. *The eigenvariables must not be free in $\overline{P}, \Gamma, \Delta$.*

(b) Examples of geometric theories. We give some examples of elementary theories that can be given a geometric axiomatization. In some cases, as in plane projective geometry (example 3), a careful choice of the basic notions is needed, or else the theory fails to be geometric. It turns out that the standard axiomatizations of projective and affine geometry use what were called **co-geometric implications** in Chapter 5, and that a geometric axiomatization is found by the use of the basic notions of intuitionistic geometry. We shall analyse further this phenomenon when presenting the duality between classical and constructive notions and proofs in the next chapter.

1. Robinson arithmetic. The language has a constant 0, a unary successor function s, binary functions $+$ and $\cdot$. Atomic formulas have the form $r = t$, for arbitrary terms r and t. For free variables, parameters $a, b, c, \ldots$ are used.

1. $\neg\, s(a) = 0$,
2. $s(a) = s(b) \supset a = b$,
3. $a = 0 \lor \exists y\, a = s(y)$,
4. $a + 0 = a$,
5. $a + s(b) = s(a + b)$,
6. $a \cdot 0 = 0$,
7. $a \cdot s(b) = a \cdot b + a$.

A classically equivalent axiomatization is obtained if 3 is replaced by:

$$3'.\ \ \neg\, a = 0 \supset \exists y\, a = s(y).$$

The axiom is not geometric because it has the implication $a = 0 \supset \bot$ in the antecedent of an implication.

2. Ordered fields and real-closed fields

I Axioms for non-degenerate linear order:
    1. $a \leqslant a$,
    2. $a \leqslant b \lor b \leqslant a$,
    3. $a \leqslant b\, \&\, b \leqslant c \supset a \leqslant c$,
    4. $\neg\, 1 \leqslant 0$.
II Axioms for ordered additive groups:
    5. $(a + b) + c = a + (b + c)$,
    6. $a + b = b + a$,
    7. $a + 0 = a$,
    8. $\exists y\, a + y = 0$,
    9. $a \leqslant b \supset a + c \leqslant b + c$.

III Axioms for multiplication:

   10. $(a \cdot b) \cdot c = a \cdot (b \cdot c)$,
   11. $a \cdot b = b \cdot a$,
   12. $a \cdot 1 = a$,
   13. $a = 0 \vee \exists y\, a \cdot y = 1$,
   14. $a \cdot (b + c) = a \cdot b + a \cdot c$,
   15. $a \leqslant b \,\&\, 0 \leqslant c \supset a \cdot c \leqslant b \cdot c$.

A classically equivalent axiomatization is obtained if in place of 13 the following axiom is chosen:

   13'. $\neg\, a = 0 \supset \exists y\, a \cdot y = 1$.

This axiom, however, is not geometric, for the same reason as above. **Real-closed fields** are obtained by adding the axioms that state the existence of square roots and zeros of polynomials of odd degree:

   16. $0 \leqslant a \supset \exists y\, a = y \cdot y$,

   17. $a_{2n+1} = 0 \vee \exists y\, a_{2n+1} \cdot y^{2n+1} + a_{2n} \cdot y^{2n} + \ldots a_1 \cdot y + a_0 = 0$.

The classically equivalent axiomatization with 17 replaced by

   17'. $\neg\, a_{2n+1} = 0 \supset \exists y\, a_{2n+1} \cdot y^{2n+1} + a_{2n} \cdot y^{2n} + \ldots a_1 \cdot y + a_0 = 0$

is clearly not geometric.

3. Classical projective geometry with constructions. The basic concepts are: equality of two points, equality of two lines, and incidence of a point on a line. These are written $a = b$, $l = m$, and $a \in l$. There are two constructions: those of a connecting line $ln(a, b)$ of two points and of an intersection point $pt(l, m)$ of two lines.

   I Axioms for the equality relations:

   $a = a, \quad a = c \,\&\, b = c \supset a = b$,

   $l = l, \quad l = n \,\&\, m = n \supset l = m$.

   II Axioms of incidence:

   $a = b \vee a \in ln(a, b), \quad a = b \vee b \in ln(a, b)$,

   $l = m \vee pt(l, m) \in l, \quad l = m \vee pt(l, m) \in m$.

   III Uniqueness axiom:

   $a \in l \,\&\, b \in l \,\&\, a \in m \,\&\, b \in m \supset a = b \vee l = m$.

IV Substitution axioms:

$$a \in l \,\&\, a = b \supset b \in l,$$

$$a \in l \,\&\, l = m \supset a \in m.$$

V Existence of three non-collinear points:

$$\exists x \exists y \exists z (\neg\, x = y \,\&\, \neg\, z \in ln(x, y)).$$

Because of axiom V, the above is not a geometric theory. We obtain a geometric axiomatization by the use of apartness between points and lines as a basic notion, instead of equality:

4. Constructive projective geometry with constructions. The basic concepts are: $a \neq b$, $l \neq m$, and $a \notin l$. The constructions are: $ln(a, b)$ and $pt(l, m)$.

I Axioms for apartness relations:

$$\neg\, a \neq a, \quad a \neq b \supset a \neq c \vee b \neq c,$$

$$\neg\, l \neq l, \quad l \neq m \supset l \neq n \vee m \neq n.$$

II Axioms of incidence:

$$\neg\, (a \neq b \,\&\, a \notin ln(a, b)),$$

$$\neg\, (a \neq b \,\&\, b \notin ln(a, b)),$$

$$\neg\, (l \neq m \,\&\, p\, t(l, m) \notin l),$$

$$\neg\, (l \neq m \,\&\, p\, t(l, m) \notin m).$$

III Uniqueness axiom:

$$a \neq b \,\&\, l \neq m \supset a \notin l \vee b \notin l \vee a \notin m \vee b \notin m.$$

IV Substitution axioms:

$$a \notin l \supset a \neq b \vee b \notin l,$$

$$a \notin l \supset l \neq m \vee a \notin m.$$

V Existence of three non-collinear points:

$$\exists x \exists y \exists z (x \neq y \,\&\, z \notin ln(x, y)).$$

As can be seen, we have now a geometric axiomatization of projective geometry. The classical basic concepts of equality and incidence are defined as negations of the basic concepts of constructive geometry, namely apartness of two points, apartness of two lines, and the 'outsideness' of a point from a line. On the other hand, these latter concepts can be defined as negations of the basic concepts of classical projective geometry, if the law of double negation is assumed: then, for example, $\neg\, a \notin l$ will be equivalent to $a \in l$.

Transitivity of equality in the classical axiomatization was given in the 'Euclidean' form, because, put in this way, it comes directly out as the contrapositive of the 'co-transitivity' of apartness.

Chapter 10 is devoted to an analysis of plane projective and affine geometry.

8.2 Proof theory of geometric theories

We shall turn after the above examples of geometric theories to the general proof theory of systems of geometric rules. In the extension of sequent calculi by such rules, all the structural properties are preserved.

Definition 8.3. *Let* **T** *be a geometric theory. Then* **G3cT** (**G3imT**) *is the system of sequent calculus obtained by adding to* **G3c** (**G3im**) *the geometric rules that correspond to the geometric axioms of* **T**, *together with the rules that arise from the closure condition.*

Theorem 8.4. Equivalence of axiomatic systems and rule systems. *A geometric axiom is derivable from the corresponding geometric rule. Conversely, a geometric rule is derivable from the corresponding geometric axiom in* **G3imT+Ctr+Cut.**

Proof. A geometric axiom $\rightarrow A$, represented as a sequent with an empty antecedent, is derivable from the corresponding geometric rule scheme. Below, an asterisk * denotes possibly iterated steps of a rule, the premisses are clearly derivable by $R\&$, and $M_i(z_i/y_i) \equiv \&\overline{Q}_i(z_i/y_i)$:

$$\dfrac{\overline{Q}_1(z_1/y_1),\overline{P} \rightarrow M_1(z_1/y_1),\ldots,M_n(z_n/y_n)}{\overline{Q}_1(z_1/y_1),\overline{P} \rightarrow \exists y_1 M_1,\ldots,\exists y_n M_n}\; R\exists^* \quad \cdots \quad \dfrac{\overline{Q}_n(z_n/y_n),\overline{P} \rightarrow M_1(z_1/y_1),\ldots,M_n(z_n/y_n)}{\overline{Q}_n(z_n/y_n),\overline{P} \rightarrow \exists y_1 M_1,\ldots,\exists y_n M_n}\; R\exists^*$$

$$\dfrac{}{\dfrac{\overline{P} \rightarrow \exists y_1 M_1,\ldots,\exists y_n M_n}{\dfrac{\overline{P} \rightarrow \exists y_1 M_1 \vee \ldots \vee \exists y_n M_n}{\dfrac{P_1\&\ldots\&P_m \rightarrow \exists y_1 M_1 \vee \ldots \vee \exists y_n M_n}{\dfrac{\rightarrow P_1\&\ldots\&P_m \supset \exists y_1 M_1 \vee \ldots \vee \exists y_n M_n}{\rightarrow \forall\overline{x}(P_1\&\ldots\&P_m \supset \exists y_1 M \vee \ldots \vee \exists y_n M_n)}\; R\forall}\; R\supset}\; L\&^*}\; R\vee^*}\; GRS$$

Conversely, a geometric rule is derivable from the corresponding geometric axiom $\rightarrow A$ in **G3im+Ctr+Cut** as shown below. In the derivation of the left premiss of cut, inverses of rules are used. These are admissible (and height-preserving) steps in **G3im**. Observe that the variable restriction in *GRS* now comes into use in the steps of $L\exists$:

$$\cfrac{\cfrac{\cfrac{\to \forall \vec{x}(P_1 \&\ldots\& P_m \supset \exists y_1 M_1 \vee \ldots \vee \exists y_n M_n)}{\cfrac{\to P_1 \&\ldots\& P_m \supset \exists y_1 M_1 \vee \ldots \vee \exists y_n M_n}{P_1 \&\ldots\& P_m \to \exists y_1 M_1 \vee \ldots \vee \exists y_n M_n}R\supset Inv} R\forall Inv}{\overline{P} \to \exists y_1 M_1 \vee \ldots \vee \exists y_n M_n}L\&Inv \qquad \cfrac{\cfrac{\cfrac{\overline{Q}_1(z_1/y_1), \overline{P}, \Gamma \to \Delta}{M_1(z_1/y_1), \overline{P}, \Gamma \to \Delta}L\&^*}{\exists y_1 M_1, \overline{P}, \Gamma \to \Delta}L\exists \quad \ldots \quad \cfrac{\cfrac{\overline{Q}_n(z_n/y_n), \overline{P}, \Gamma \to \Delta}{M_n(z_n/y_n), \overline{P}, \Gamma \to \Delta}L\&^*}{\exists y_n M_n, \overline{P}, \Gamma \to \Delta}L\exists}{\exists y_1 M_1 \vee \ldots \vee \exists y_n M_n, \overline{P}, \Gamma \to \Delta}L\vee^*}{\cfrac{\overline{P}, \overline{P}, \Gamma \to \Delta}{\overline{P}, \Gamma \to \Delta}Ctr^*}Cut$$

QED.

Remark. It is clear by the above derivation that the geometric rule scheme hides a cut on the formula $\exists y_1 M_1 \vee \ldots \vee \exists y_n M_n$. The substituted variables y_i are bound variables in the virtual cut formula, so it will be convenient to regard the substituted variables of the scheme as bound variables and to assume that in a derivation the sets of free and bound variables are disjoint.

The inversion lemmas for the propositional rules that hold for **G3c** and **G3im** hold for their geometric extensions as well, because the geometric rule scheme has only atomic formulas as principal and active formulas. So we have:

Lemma 8.5. Inversion. *All the inversions of the propositional rules that hold for **G3c** and **G3im** hold also for their geometric extensions.*

For the inversions of $L\exists$ and $R\forall$, we need to add a condition on the variable to avoid clashes with the proper variables of the geometric rules in the derivation.

Lemma 8.6. Substitution. *Given a derivation of $\Gamma \to \Delta$ in* **G3cT** *(**G3imT**), with x a free variable in Γ, Δ, t a term free for x in Γ, Δ and not containing any of the variables of the geometric rules in the derivation, we can find a derivation of $\Gamma(t/x) \to \Delta(t/x)$ in* **G3cT** *(**G3im**) with the same height.*

Proof. By induction on the height of the given derivation. For the logical rules, the proof is contained in lemma 4.1.2 of *Structural Proof Theory*, so we need to consider only the cases that arise from the addition of the geometric rule scheme. Suppose the last rule in the derivation of $\Gamma \to \Delta$ is *GRS*, with premisses

$$\overline{Q}_i(z_i/y_i), \overline{P}, \Gamma' \to \Delta$$

for $i = 1, \ldots, n$. The $\overline{Q}_i$ are atomic and the term t is free for x in these premisses; thus by the induction hypothesis we get derivations of

$$\overline{Q}_i(z_i/y_i)(t/x), \overline{P}(t/x), \Gamma'(t/x) \to \Delta(t/x)$$

Since x is a free variable in Γ, Δ, by the remark at the end of the previous section we have $x \neq y_i$, and since the z_i are not free in $\overline{P}, \Gamma, \Delta$ we have $x \neq z_i$. Moreover, by hypothesis, t does not contain any of the y_i. Therefore the two substitutions in $\overline{Q}_i$ are independent and we have $Q_i(z_i/y_i)(t/x) \equiv Q_i(t/x)(z_i/y_i)$. Since t does not contain any of the z_i, the z_i are not free in $\overline{P}(t/x), \Gamma'(t/x) \to \Delta(t/x)$, so we can apply the geometric rule scheme to the premisses

$$\overline{Q}_i(t/x)(z_i/y_i), \overline{P}(t/x), \Gamma'(t/x) \to \Delta(t/x)$$

Now we get $\overline{P}(t/x), \Gamma'(t/x) \to \Delta(t/x)$, i.e., $\Gamma(t/x) \to \Delta(t/x)$.   QED.

Lemma 8.7. Inversion for quantifier rules.

(i) If $\vdash_n \exists x A, \Gamma \to \Delta$ and y is not among the variables of the geometric rules in the derivation, then $\vdash_n A(y/x), \Gamma \to \Delta$.

(ii) If $\vdash_n \Gamma \to \Delta, \forall x A$ and y is not among the variables of the geometric rules in the derivation, then $\vdash_n \Gamma \to \Delta, A(y/x)$.

Proof. (i) By induction on height of derivation n, as in Lemma 6.2. If $n = 0$, then $\exists x A, \Gamma \to \Delta$ is either an initial sequent, or a conclusion of $L \bot$, or a conclusion of a zero-premiss geometric rule. In each case also $A(y/x), \Gamma \to \Delta$ is an initial sequent, or a conclusion of $L \bot$, or a conclusion of a zero-premiss geometric rule, thus $\vdash_0 A(y/x), \Gamma \to \Delta$.

If $n > 0$ and $\exists x A$ is principal in the last rule, the premiss gives a derivation of $A(z/x), \Gamma \to \Delta$, where z is not free in Γ, Δ. By Lemma 8.6, using the substitution y/z, we obtain a derivation of the same height of $A(y/x), \Gamma \to \Delta$. If $\exists x A$ is not principal in the last rule, we argue as in lemma 4.2.3 of *Structural Proof Theory* in the case the last rule is a logical rule. If the last rule is a geometric rule, with $\Gamma \equiv \overline{P}, \Gamma'$ and premisses

$$\overline{Q}_1(z_1/y_1), \overline{P}, \exists x A, \Gamma' \to \Delta, \ldots, \overline{Q}_n(z_n/y_n), \overline{P}, \exists x A, \Gamma' \to \Delta$$

we get by the assumption that free and bound variables are disjoint and by the inductive hypothesis derivations of the sequents

$$\overline{Q}_1(z_1/y_1), \overline{P}, A(y/x), \Gamma' \to \Delta, \ldots, \overline{Q}_n(z_n/y_n), \overline{P}, A(y/x), \Gamma' \to \Delta$$

Since y is none of the z_i, we can apply the geometric rule scheme to these premisses and obtain a derivation of $\overline{P}, A(y/x), \Gamma' \to \Delta$.

(ii) Similar to (i). QED.

Without loss of generality, we can assume by Lemma 8.6 that the following condition on variables is satisfied:

Disjointness condition. *In a derivation in* **G3cT** (**G3imT**), *the collections of proper variables of any two geometric rules are disjoint.*

Theorem 8.8. *The rules of weakening*

$$\frac{\Gamma \to \Delta}{A, \Gamma \to \Delta} LW \qquad \frac{\Gamma \to \Delta}{\Gamma \to \Delta, A} RW$$

are height-preserving admissible in **G3cT** *and in* **G3imT**.

Proof. Consider the last step in the derivation of the premiss of the weakening rules. Apply, inductively, weakening to the premisses $\Gamma_i \to \Delta_i$ of the last step to obtain $A, \Gamma_i \to \Delta_i$, then the rule that had been used as the last step, which gives $A, \Gamma \to \Delta$. Observe that if the last rule is geometric and the weakening formula A contains some of its variables, the variable condition is no longer satisfied after weakening with A. In this case the substitution lemma is applied to the premisses of the geometric rule to have new free variables that do not clash with those in A. The conclusion is then obtained by applying the inductive hypothesis and the geometric rule. QED.

Theorem 8.9. *The rules of contraction*

$$\frac{A, A, \Gamma \to \Delta}{A, \Gamma \to \Delta} LC \qquad \frac{\Gamma \to \Delta, A, A}{\Gamma \to \Delta, A} RC$$

are height-preserving admissible in **G3cT** *and in* **G3imT**.

Proof. By simultaneous induction for left and right contraction on the height of the derivation of the premiss. If it is an initial sequent, also the conclusion is. If the last rule is a propositional rule, then the conclusion follows as in theorem 3.2.2 of *Structural Proof Theory*. If it is $L\forall$, we apply the induction hypothesis to the premiss of the rule, and then the rule, and similarly if it is $L\exists$ with A not principal in it. If it is $L\exists$ with $A \equiv \exists x B$ and premiss $B(y/x), \exists x B, \Gamma \to \Delta$, by the variable condition on the geometric rule and the Remark after Theorem 8.4, y is not a variable in any geometric rule in the derivation, so we can apply the inversion lemma for $L\exists$ instantiated by y and obtain a derivation of $B(y/x), B(y/x), \Gamma \to \Delta$. By the induction hypothesis we get $B(y/x), \Gamma \to \Delta$ and by $L\exists, \exists x B, \Gamma \to \Delta$.

If the last rule is a geometric rule, we distinguish three cases: 1. No occurrence of A is principal in the rule. 2. One occurrence of A is principal, the other is not. 3. Both occurrences of A are principal.

The first case is handled by a straightforward induction and the second by the repetition of the principal formulas P_i in the premisses of the geometric rule. Finally, the third case is taken care of by the closure condition.   QED.

We are now in the position to prove the admissibility of cut for our rule systems for geometric theories. We remark that the proof has the same structure as the proof of admissibility of cut for universal theories (Theorem 6.9 above), with an additional use of the substitution lemma in order to meet the variable restriction in the geometric rule scheme.

Theorem 8.10. *The rule of cut*

$$\frac{\Gamma \to \Delta, A \qquad A, \Gamma' \to \Delta'}{\Gamma, \Gamma' \to \Delta, \Delta'} \; Cut$$

is admissible in **G3cT** *and in* **G3imT**.

Proof. By induction on the length of A with a subinduction on the sum of the heights of the derivations of $\Gamma \to \Delta, A$ and $A, \Gamma' \to \Delta'$. We need to consider only the cases that arise from the addition of the geometric rule scheme. The other cases are treated in the corresponding proof for **G3c** and **G3im**.

1. If the left premiss is a zero-premiss geometric rule, then also the conclusion is a zero-premiss geometric rule, because these have an arbitrary context as succedent.

2. If the right premiss is a zero-premiss geometric rule with A not principal in it, the conclusion is a zero-premiss geometric rule for the same reason as in case *1*.

3. If the right premiss is a zero-premiss geometric rule with A principal in it, A is atomic and we consider the left premiss. The case that it is a geometric zero-premiss rule is covered by *1*. If it is an initial sequent with A not principal, the conclusion is a logical axiom; otherwise, Γ contains the atom A and the conclusion follows from the right premiss by weakening. In the remaining cases we consider the last rule in the derivation of $\Gamma \to \Delta, A$. Since A is atomic, A is not principal in the rule. Let us consider the case of

a geometric rule (the others being dealt with similarly, except $R \supset$ and $R \forall$ that are covered in 4). The derivation ends with

$$\dfrac{\dfrac{\overline{Q}_1(z_1/y_1), \overline{P}, \Gamma'' \to \Delta, A \quad \ldots \quad \overline{Q}_n(z_n/y_n), \overline{P}, \Gamma'' \to \Delta, A}{\overline{P}, \Gamma'' \to \Delta, A} \, {}_{GRS} \qquad A, \Gamma' \to \Delta'}{\overline{P}, \Gamma', \Gamma'' \to \Delta, \Delta'} \, Cut$$

The cut cannot be simply permuted to the premisses of *GRS* because Γ', Δ' may bring in free variables that clash with the proper variables z_i and thus prevent the application of *GRS* after the cuts. Instead, we apply first the substitution lemma to the right premiss of cut and replace all the variables z_i, if any, by fresh variables w_i, and denote the substitution by w/z. Observe that by the variable condition in *GRS*, the substitution does not affect the cut formula A. We obtain the sequent

$$A, \Gamma'(w/z) \to \Delta'(w/z)$$

Now n cuts with the n premisses of *GRS* give, for $i = i, \ldots n$, the n sequents

$$\overline{Q}_i(z_i/y_i), \overline{P}, \Gamma'(w/z), \Gamma'' \to \Delta, \Delta'(w/z)$$

By applying *GRS* to these n premisses we get

$$\overline{P}, \Gamma'(w/z), \Gamma'' \to \Delta, \Delta'(w/z)$$

The derivation is continued as before with the substitution (w/z) performed globally. Observe that by the disjointness condition, the substitution does not affect the active formulas of other geometric rules in the derivation. The cut has thus been replaced by n cuts with a left premiss that has a derivation of lower height and a right premiss of unaltered height.

Let us now consider the cases in which neither premiss is an axiom.

4. A is not principal in the left premiss. These cases are dealt with as above, with cut permuted upwards to the premisses of the last rule used in the derivation of the left premiss (with a suitable variable substitution to match the variable restrictions in the cases of quantifier rules and the geometric rule-scheme), except for $R \supset$ and $R \forall$ in **G3imT**. By the intuitionistic restrictions in these rules, A does not appear in the premisses, and the conclusion is obtained without cut by $R \supset$ ($R \forall$, resp.) and weakening.

5. A is principal only in the left premiss. Then A has to be a compound formula. Therefore, if the last rule of the right premiss is a geometric rule, A

cannot be principal in the rule, because only atomic formulas are principal in geometric rules. In this case cut is permuted to the premisses of the right premiss, with appropriate substitution of free variables as in 3, if the right premiss is a *GRS*. If the right rule is a logical one with *A* not principal in it, the usual reductions are applied.

6. *A* is principal in both premisses. This case can involve only logical rules and is dealt with as in the usual proof for pure logic.   QED.

Examples of proof analyses in geometric theories will be given in Chapter 10. In typical cases, it turns out that existential axioms are conservative relative to the derivability of atomic cases from atomic assumptions. If for the rest of the rules a subterm property can be shown, a positive solution for the said derivability problem follows. In this chapter, we shall present instead a general result about the proof theory of geometric theories.

8.3 Barr's theorem

We apply here the method of extension by rules to a general result for geometric theories.

Barr's theorem. *Let T be a geometric theory and let A be a geometric implication. If* **G3cT** $\vdash \rightarrow A$, *then* **G3imT** $\vdash \rightarrow A$.

Proof. Let *A* be $\forall \overline{x}(B \supset C)$ and consider a proof of *A* in **G3cT**. Because *B* and *C* do not contain $\supset$ or $\forall$, the derivation of *A* consists of geometric rules, rules for &, $\vee$, $\exists$, a step of $R\supset$ and steps of $R\forall$. The geometric rules can occur in any order among the logical rules, however, of the logical rules, $R\supset$ and $R\forall$ come last. The geometric rules have the same succedent in both the premisses and the conclusion, so therefore rules $R\supset$ and $R\forall$ are applied to single-succedent sequents. It follows that the given proof must already be a proof in **G3imT**.   QED.

Therefore, through the method of proof analysis, the result reduces to a triviality: A classical proof of a geometric implication in a geometric theory formulated as a sequent system with rules is already an intuitionistic proof. If we add the requirement that the geometric implication must not contain $\perp$ in the antecedent, then the classical proof is even a proof in minimal logic.

Notes to Chapter 8

Interest in the study of geometric theories has arisen from different areas of logic and mathematics. In topos theory, geometric formulas are characterized as the fragment of first-order logic preserved by geometric morphisms. The statement of Barr's theorem in that context is as follows: *For any Grothendieck topos* **E**, *there is a Boolean topos* **B** *and a surjective geometric morphism* $\gamma : \mathbf{B} \to \mathbf{E}$. Thus the conservativity result for geometric theories is proved using a completeness theorem for geometric theories in Grothendieck topoi and the construction of a suitable Boolean topos out of a Grothendieck topos (cf. Johnstone 1977, Mac Lane and Moerdijk 1992, p. 513). The preservation property extends beyond first-order logic to existential fixed-point formulas (cf. Blass 1988).

Palmgren (2002) indicates a proof-theoretical proof of Barr's theorem by showing that geometric implications are stable under the Dragalin-Friedman translation. A proof for the special case of the empty geometric theory is suggested in Troelstra and Van Dalen (1988) (exercise 2.6.14) by means of Kripke models.

The nature of geometric logic as the logic of finite observations has been emphasized in localic approaches to constructive topology (cf. Vickers 1989). Geometric theories can be treated, as any other theory, by the addition of Hilbert-style axioms to a logical proof system, but axiomatic systems are hard to analyse proof-theoretically. In Simpson (1994), geometric theories are presented through suitable rules that extend intuitionistic natural deduction. A proof of normalization for the extensions thus obtained is given and the systems applied in a systematic study of the proof theory of intuitionistic modal logic. Finally, in Coste, Lombardi, and Roy (2001), the so-called method of 'dynamical proof', which establishes the derivability of one atom from a finite set of atoms, is applied to certain specific geometric theories such as the theory of algebraically closed fields.

We observe that Barr's theorem is not a characterization of the intuitionistic fragment of geometric theories, because we can go beyond geometric implications and maintain the conservativity result. First, following Dragalin's suggestion (cf. Section 3.7.3 in Troelstra and Schwichtenberg 2000), we can modify the intuitionistic left rule for implication by admitting a multisuccedent conclusion in the left premiss

$$\frac{A \supset B, \Gamma \to \Delta, A \quad B, \Gamma \to \Delta}{A \supset B, \Gamma \to \Delta} \; L\supset$$

Rule $L\supset$ of the classical calculus, without $A \supset B$ in the left premiss, is then admissible in the modified intuitionistic calculus; thus the difference between the intuitionistic and classical sequent systems is confined to rules $R\supset$ and $R\forall$. An operational definition of formulas for which the conservativity of classical derivations holds can be given: if a formula is derivable classically in a geometric theory and the derivation contains no steps of $R\supset$ and $R\forall$ with a non-empty context in the premiss, then the derivation is an intuitionistic derivation. However, this is an empty characterization that states nothing but that 'an intuitionistic derivation is an intuitionistic derivation'. A characterization in terms of only the form of the formulas, not of their derivations, would be desirable. There are classes of formulas, such as geometric implications, the form of which forces the derivation to be of the stated kind. The same is true, for example, if the formula does not contain in its positive part implications or universally quantified formulas as components of a disjunction. Even so, there are still formulas outside the mentioned classes for which the conservativity holds.

9 | Classical and intuitionistic axiomatics

The method of conversion of mathematical axioms into rules of sequent calculus reveals a perfect duality between classical and constructive basic notions, such as equality and apartness, and between the respective rules for these notions. Derivations by the mathematical rules of a constructive theory are mirror-image duals of corresponding classical derivations, the mathematical rules being obtained by shifting from the left to the right rule scheme and *vice versa*.

The class of geometric theories is among those convertible into rules and the duality defines the class of co-geometric theories, as in Definition 5.2. The logical rules of classical sequent calculus are invertible, which has for quantifier-free theories the effect that logical rules in derivations can be permuted to apply after the mathematical rules. In the case of mathematical rules that have variable conditions, this separation of logic does not always hold, because quantifier rules may fail to permute down. A sufficient condition for the permutability of mathematical rules is determined in this chapter and applied to give an extension of Herbrand's theorem from universal to geometric and co-geometric theories.

The use of systems of left and right rules is a matter of choice. In Section 7.2, we used a system of right rules for linear order, because it was felt easier to prove the main results. These results can be proved through a system of left rules, as well.

9.1 The duality of classical and constructive notions and proofs

(a) **Finitary basic concepts.** A constructive approach to the real numbers uses the apartness of two real numbers as a basic relation. The axioms for this relation, written $a \neq b$, are as follows:

> AP1 $\neg\, a \neq a$,
>
> AP2 $a \neq b \supset a \neq c \vee b \neq c$.

Substituting a for c in AP2, we get $a \neq b \supset a \neq a \vee b \neq a$, so symmetry of apartment follows by AP1. Equality is a defined notion:

EQDEF $a = b \equiv \neg\, a \neq b$.

By AP1, equality is reflexive. By the contraposition of symmetry of apartness, we have also symmetry of equality. By AP2 and symmetry of apartness, we have $a \neq b \supset a \neq c \vee c \neq b$, so contraposition gives transitivity of equality.

If instead of the constructively motivated notion of apartness we take equality as a basic notion, with its standards properties of reflexivity, symmetry, and transitivity, apartness can be defined by

APDEF $a \neq b \equiv \neg\, a = b$.

Irreflexivity and symmetry of apartness follow. For the 'splitting' property of an apartness $a \neq b$ into two cases $a \neq c \vee b \neq c$, the contraposition of transitivity of equality gives $\neg\, a = b \supset \neg\,(a = c \,\&\, c = b)$. To distribute negation inside the conjunction, classical logic is needed.

The play with classical and constructive notions can be carried further in geometry. The parallelism of two lines is a classical basic relation, and its constructive counterpart is the 'convergence' of two lines l and m, written $l \nparallel m$. The axioms are as for the apartness relation above.

The intuition for constructive basic notions is that the classical notions such as equality are 'infinitely precise,' whereas apartness, if it obtains, can be verified by a finite computation. Something of this intuition can be seen already in Brouwer's first ideas on the topic of apartness relations, from 1924, where it is required that the set of objects considered be continuous. This was certainly the intention with Brouwer's constructive real numbers and with Heyting's constructive synthetic geometry. A set is defined as discrete if it has a decidable equality relation, otherwise it is continuous. The constructive interpretation of the law of excluded middle for equality, $a = b \vee \neg\, a = b$, is precisely that the basic domain of objects is discrete. With such sets, it makes no difference which relations are used as basic, the constructive or classical ones, as the axioms are interderivable.

In von Plato (2001c), the constructivization of elementary axiomatics was extended from order relations to lattice theory. It then seemed that proofs that use apartness relations would be harder to find than corresponding classical proofs (see especially theorem 7.1 and the discussion on p. 196). It has turned out, however, that there is an automatic bridge between classical and constructive notions and proofs. The matter is best seen on a formal level if for the representation of proofs sequent calculus is used.

(b) Derivations in left and right rule systems. We shall show the duality of classical and constructive notions and proofs through examples that are easily seen to be representative of the general situation. Consider the theory of apartness. Its two axioms convert into the system of left rules

$$\frac{}{a \neq a, \Gamma \to \Delta}\,\mathit{Irref} \qquad \frac{a \neq c, a \neq b, \Gamma \to \Delta \quad b \neq c, a \neq b, \Gamma \to \Delta}{a \neq b, \Gamma \to \Delta}\,\mathit{Split}$$

Symmetry of apartness is expressed by the sequent $\to a \neq b \supset b \neq a$ and has the derivation

$$\frac{\dfrac{\dfrac{}{a \neq a, a \neq b \to b \neq a}\,\mathit{Irref} \quad b \neq a, a \neq b \to b \neq a}{a \neq b \to b \neq a}\,\mathit{Split}}{\to a \neq b \supset b \neq a}\,R\supset \tag{1}$$

Now take rules *Irref* and *Split* and move all atoms to the other side by rule $R\neg$ of classical sequent calculus. Next write $a = b$ for $\neg a \neq b$, etc. The result can be written as the two rules for equality:

$$\frac{}{\Gamma \to \Delta, a = a}\,\mathit{Ref} \qquad \frac{\Gamma \to \Delta, a = b, a = c \quad \Gamma \to \Delta, a = b, b = c}{\Gamma \to \Delta, a = b}\,\mathit{ETr}$$

Here *ETr* stands for 'Euclidean transitivity', from the way transitivity is expressed by Euclid.

Within our example derivation (1), switch atoms on the left and right sides of the arrow, erase the slashes, and change the rule names to get

$$\frac{\dfrac{\dfrac{}{b = a \to a = b, a = a}\,\mathit{Ref} \quad b = a \to a = b, b = a}{b = a \to a = b}\,\mathit{ETr}}{\to b = a \supset a = b}\,R\supset \tag{2}$$

The sequents in the mathematical part of the derivation (2) are perfect mirror images of those in derivation (1).

Next we convert the two axioms of an apartness relation into a system of right rules:

$$\frac{\Gamma \to \Delta, a \neq a}{\Gamma \to \Delta}\,\mathit{Irref} \qquad \frac{\Gamma \to \Delta, a \neq c, b \neq c, a \neq b}{\Gamma \to \Delta, a \neq c, a \neq b}\,\mathit{Split}$$

The symmetry of apartness now has the derivation

$$\frac{\dfrac{\dfrac{a \neq b \to b \neq a, a \neq a, a \neq b}{a \neq b \to b \neq a, a \neq a}\,\mathit{Split}}{a \neq b \to b \neq a}\,\mathit{Irref}}{\to a \neq b \supset b \neq a}\,R\supset \tag{3}$$

The mirror image left rules for equality are

$$\frac{a = a, \Gamma \to \Delta}{\Gamma \to \Delta} \, Ref \qquad \frac{a = b, a = c, b = c, \Gamma \to \Delta}{a = c, b = c, \Gamma \to \Delta} \, ETr$$

Symmetry is derived by the mirror image of derivation (3):

$$\frac{\dfrac{\dfrac{\dfrac{a = b, a = a, b = a \to a = b}{a = a, b = a \to a = b} \, Etr}{b = a \to a = b} \, Ref}{\to b = a \supset a = b} \, R\supset} \tag{4}$$

There are thus two kinds of systems of rules of equality, and the same for apartness. Euclidean equality has axioms that are Harrop formulas. As a consequence, derivations by the two rules of this theory are linear, with just one premiss. Also the mirror image right theory of apartness has linear derivations. It could be called a 'co-Harrop' theory, with axioms that have no conjunctions in their negative parts.

The above examples of rules and derivations are fully representative of the general situation: we can take the rule scheme that acts on the left, antecedent part of sequents, and convert it into a rule scheme that acts on the right, succedent part, in exactly the same way as in the examples. There will be a change in the basic notions, from constructive to classical or the other way around. The question remains what, if anything, is gained by the constructivization of classical elementary axiomatic theories; combinatorially, for each derivation in a constructive system of rules, there is a dual classical derivation and *vice versa*.

9.2 From geometric to co-geometric axioms and rules

We introduced in Chapter 8 a sequent calculus formulation of geometric theories and presented as examples of geometric theories real-closed fields, Robinson arithmetic, and constructive projective geometry. As noted, to obtain a geometric axiomatization, some care is needed when formulating the axioms: for example, the axiom that states the existence of inverses on non-zero elements in the theory of fields is

$$\neg x = 0 \supset \exists y \, x \cdot y = 1$$

This axiom is not geometric, because it contains an implication the antecedent of which is an implication ($x = 0 \supset \bot$), but it can be replaced by the geometric axiom

$$a = 0 \vee \exists y \; a \cdot y = 1$$

In this formulation, the axiom can be converted into a rule that follows the geometric rule scheme:

$$\frac{a = 0, \Gamma \rightarrow \Delta \quad a \cdot y = 1, \Gamma \rightarrow \Delta}{\Gamma \rightarrow \Delta} \; \textit{L-inv}$$

The rule has the variable condition that y be not free in Γ, Δ.

Alternatively, we can take apartness $a \neq b$ as the primitive relation and turn rule *L-inv* into the following right rule, with the same variable condition on y:

$$\frac{\Gamma \rightarrow \Delta, a \neq 0 \quad \Gamma \rightarrow \Delta, a \cdot y \neq 1}{\Gamma \rightarrow \Delta} \; \textit{R-inv}$$

This form of the rule corresponds to the axiom $\neg \forall y (a \neq 0 \; \& \; a \cdot y \neq 1)$.

All the other axioms for fields and real-closed fields can be given in terms of right rules for the primitive relation of apartness.

A similar transformation can be made with the axioms of constructive affine geometry. These axioms, presented in von Plato (1995), are based on the primitive notions of distinct points $a \neq b$, distinct lines $l \neq m$, convergent lines $l \nparallel m$, and of a point outside a line $a \notin l$, and on the constructions of a line $ln(a, b)$ that connects two distinct points a and b and of a point $pt(l, m)$ obtained as the intersection of two convergent lines l and m.

In 8.1(b), we observed that the theory extended by the axiom that states the existence of three non-collinear points remains geometric:

$$\exists x \exists y \exists z (x \neq y \; \& \; z \notin ln(x, y))$$

The axiom corresponds to the following instance of the geometric rule scheme:

$$\frac{x \neq y, z \notin l(x, y), \Gamma \rightarrow \Delta}{\Gamma \rightarrow \Delta}$$

The variable condition is that x, y, z must not occur free in Γ, Δ.

If the axiomatization is instead based on the primitive relations of equality of points, equality of lines, parallelism of lines, and incidence of a point with a line, the axiom becomes:

$$\exists x \exists y \exists z (\neg x = y \; \& \; \neg z \in ln(x, y))$$

This axiom is not any longer a geometric implication, so we have to make, somewhat annoyingly, the following Remark:

Remark. *Classical geometry is not a geometric theory.*

The axiom can be given, however, in the form of the following right rule, with the condition that x, y, z must not occur free in Γ, Δ:

$$\frac{\Gamma \to \Delta, x = y, z \in l(x, y)}{\Gamma \to \Delta}$$

by which

$$\neg \forall x \forall y \forall z (x = y \lor z \in ln(x, y))$$

is derivable. All the other axioms can also be uniformly presented as right rules for the primitive relations $a = b$, $l = m$, $l \parallel m$, and $a \in l$.

The above examples illustrate a general result:

Theorem 9.1. *Let* **T** *be a geometric theory based on the primitive relations* R_i, *with rules that follow the geometric rule scheme* GRS, *and let* **T**′ *be the theory obtained by formulating the axioms in terms of the dual relations* R_i'. *Then a contraction- and cut-free system for the theory* **T**′ *is obtained by turning all the instances of* GRS *into the form*

$$\frac{\Gamma \to \Delta, \overline{P}', \overline{Q}_1'(\overline{z}_1/\overline{y}_1) \quad \cdots \quad \Gamma \to \Delta, \overline{P}', \overline{Q}_n'(\overline{z}_n/\overline{y}_n)}{\Gamma \to \Delta, \overline{P}'} \text{ co-GRS}$$

with the eigenvariables $\overline{z}_i$ *not free in* Γ, Δ, $\overline{P}'$ *and the apices indicating the atoms transformed in terms of the dual relations* R_i'.

We can ask what kinds of axioms are captured by the scheme co-GRS. Clearly, the scheme is interderivable with an axiom of the form

$$\forall \overline{x}(\forall \overline{y}_1 M_1' \& \ldots \& \forall \overline{y}_n M_n' \supset P_1' \lor \ldots \lor P_m') \qquad co\text{-}GA$$

in which $M_j' \equiv Q_{j_1}' \lor \ldots \lor Q_{j_{k_j}}'$.

It is easy to verify that any formula of the form

$$\forall \overline{x}(A \supset B)$$

with A and B formulas that do not contain $\supset$ or $\exists$, can be brought to a canonical form that consists of conjunctions of formulas of the form given by co-GA. We recall from Section 5.1 that formulas A, B that do not contain $\supset$ or $\exists$, are called **co-geometric** and the implication $A \supset B$ a **co-geometric**

implication. A theory axiomatized by co-geometric implications will be called a **co-geometric theory**. Classical projective and affine geometry with the axiom of non-collinearity constitute examples of co-geometric theories.

The above examples have shown how the duality between geometric and co-geometric theories can be used for changing the primitive notions in the sequent formulation of a theory. Metatheoretical results can be imported from one theory to its dual by exploiting the symmetry of their associated sequent calculi.

In Section 6.4, an extension of Herbrand's theorem to universal theories was presented (Theorem 6.19). Clearly, the theorem does not extend to geometric theories. In fact, if $\exists x P$ is an axiom of a theory T, then $\rightarrow \exists x P$ is derivable in **G3cT** but there is no finite disjunction such that $\rightarrow P(t_1) \vee \ldots \vee P(t_n)$ would be derivable in **G3cT**.

The crucial component in the proof of Herbrand's theorem is that one can assume a derivation in which the quantifier rules come last. In first-order logic and in universal theories this is unproblematic. With mathematical rules that have variable conditions, like the geometric or the co-geometric rule scheme, the quantifier rules cannot in general be permuted last in a derivation. Suppose we have a derivation with the steps

$$
\cfrac{
  \cfrac{\overline{Q}_1(z_1/y_1), \overline{P}, \Gamma \rightarrow \Delta, \exists x A, A(t/x)}{\overline{Q}_1(z_1/y_1), \overline{P}, \Gamma \rightarrow \Delta, \exists x A} \scriptstyle R\exists \qquad \ldots \qquad \overline{Q}_n(z_n/y_n), \overline{P}, \Gamma \rightarrow \Delta, \exists x A
}{\overline{P}, \Gamma \rightarrow \Delta, \exists x A} \scriptstyle GRS
$$

If term t contains the variable z_1, the permutation of $R\exists$ to below GRS fails because the variable condition for a correct application of GRS would no longer be satisfied. This is the exact structural reason for the failure of Herbrand's theorem for existential theories. We can nevertheless impose an additional hypothesis that makes the permutation possible. The hypothesis ensures that a fresh variable substitution, limited to the atoms $\overline{Q}_1$, is possible.

Lemma 9.2. *Let* **T** *be a geometric theory and let* **G3cT** *be the sequent system obtained by turning the theory into a system of left rules. Suppose that the sequent* $\overline{Q}_i(z_i/y_i), \overline{P}, \Gamma \rightarrow \Delta, A(t/x)$ *is derivable in* **G3cT**, *that* z_i *is not free in* $\overline{P}, \Gamma, \Delta$, *and that no atom* $\overline{Q}_i$ *occurs positively in* A. *Then* $\overline{Q}_i(w/y_i), \overline{P}, \Gamma \rightarrow \Delta, A(t/x)$ *is derivable for an arbitrary fresh variable* w.

Proof. Consider the initial sequents in a derivation of the given sequent. By the assumptions that z_i not occur free in Γ, Δ and that no atom among the $\overline{Q}_i$ be in a positive part of A, it follows that the principal atoms of the initial sequents are not among the $\overline{Q}_i$. Thus, after the substitution of the variable z_i

with a fresh variable w in the atoms $\overline{Q}_i(z_i/y_i)$, the leaves of the tree remain initial sequents, and the logical steps remain correct because the atoms in $\overline{Q}_i$ are never principal in logical rules. Since w is a fresh variable, also the instances of the geometric rule scheme remain correct, thus the substitution produces a derivation of $\overline{Q}_i(w/y_i), \overline{P}, \Gamma \rightarrow \Delta, A(t/x)$ in **G3cT**.   QED.

By the lemma, we can assume a derivation in which the mathematical rules come first, followed by propositional rules, followed by a linear part that consists of quantifier rules. The rest of the proof of Herbrand's theorem is then a routine matter. Thus we have:

Theorem 9.3. Herbrand's theorem for geometric theories. *Let* **T** *be a geometric theory and let* **G3cT** *be the sequent system obtained by turning the theory into a system of rules that follow the geometric rule scheme GRS. If the sequent* $\rightarrow \forall x \exists y_1 \ldots \exists y_k A$, *with A quantifier free, is derivable in* **G3cT** *and no atom* $\overline{Q}_i$ *occurs positively in A, then there are terms* t_{i_j} *with* $i \leqslant n, j \leqslant k$ *such that*

$$\rightarrow \bigvee_{i=1}^{n} A(t_{i_1}/y_1, \ldots, t_{i_k}/y_k)$$

is derivable in **G3cT**.

By exploiting the symmetry between a left and a right rule system we obtain the corresponding results for co-geometric theories.

Lemma 9.4. *Let* **T** *be a co-geometric theory and let* **G3cT** *be the sequent system obtained by turning the theory into a system of right mathematical rules. Suppose the sequent* $\Gamma \rightarrow \Delta, \overline{Q}_i(z_i/y_i), \overline{P}, A(t/x)$ *is derivable in* **G3cT**, z_i *is not free in* $\overline{P}, \Gamma, \Delta$, *and no atom* $\overline{Q}_i$ *occurs negatively in A. Then* $\Gamma \rightarrow \Delta, \overline{Q}_i(w/y_i), \overline{P}, A(t/x)$ *is derivable for an arbitrary fresh variable w.*

Theorem 9.5. Herbrand's theorem for co-geometric theories. *Let* **T** *be a co-geometric theory and let* **G3cT** *be the sequent system obtained by turning the theory into a system of right rules that follow the co-geometric rule scheme co-GRS. If the sequent* $\rightarrow \forall x \exists y_1 \ldots \exists y_k A$, *with A quantifier free, is derivable in* **G3cT** *and no atom* $\overline{Q}_i$ *occurs negatively in A, then there are terms* t_{i_j} *with* $i \leqslant n, j \leqslant k$ *such that*

$$\rightarrow \bigvee_{i=1}^{n} A(t_{i_1}/y_1, \ldots, t_{i_k}/y_k)$$

is derivable in **G3cT**.

9.3 Duality of dependent types and degenerate cases

The axiomatization of elementary geometry with constructive basic notions leads in a natural way to **dependent typing**: a formula with a constructed line $ln(a, b)$, such as the incidence axiom $a \in ln(a, b)$, is well-formed only if the condition of non-degeneracy $a \neq b$ is satisfied. In a first-order formulation, incidence axioms with conditions of non-degeneracy can be given as implications, as in von Plato (1995). For projective geometry, we have

$$a \neq b \supset \neg a \notin ln(a, b), \quad a \neq b \supset \neg b \notin ln(a, b),$$

and similarly for intersection points. The corresponding left rule for the first axiom is the zero-premiss rule

$$\frac{}{a \neq b, a \notin ln(a, b), \Gamma \to \Delta} \, Inc$$

By the duality of left and right rules, we have for the classical notions of equality and incidence the rule

$$\frac{}{\Gamma \to \Delta, a = b, a \in ln(a, b)} \, Inc$$

Thus, the incidence axioms for connecting lines in a classical formulation are

$$a = b \vee a \in ln(a, b), \quad a = b \vee b \in ln(a, b),$$

and similarly for the rest of the incidence axioms. The **degenerate cases** of equalities $a = b$ in these axioms are the classical duals of dependent typings in constructive geometry. The phenomenon is quite general; similar observations could be made about the condition for the inverse operation.

The use of constructions seems to be necessary for the conversion of mathematical axioms into systems of cut-free rules, be it a system based on classical or constructive notions. To see why, we formulate elementary geometry as a **relational theory** with existential axioms in place of constructions, as in

$$\forall x \forall y \exists z (x \in z \,\&\, y \in z).$$

(The sorts of the variables are determined from their places in the incidence relation: x and y points, z a line.) Next, uniqueness axioms are added, such as

$$\forall x \forall y \forall z \forall v (x \in z \,\&\, y \in z \,\&\, x \in v \,\&\, y \in v \supset z = v).$$

As mentioned above, it is possible to formulate geometry, the axiom of non-collinearity included, either as a constructive geometric theory, or as a classical co-geometric theory. This result refers to a formulation with geometric constructions. With a relational formulation, a comparison of the form of the existential axioms that replace constructions with the form of the axiom of non-collinearity leads instead to the following insight:

Indispensability of constructions. *If non-collinearity is formulated as a geometric implication, the existence axioms are co-geometric; if instead the existence axioms are geometric, non-collinearity is co-geometric.*

There is thus a fundamental incompatibility in both approaches, but it can be overcome through the use of constructions. This phenomenon is quite general and is met in, for example, lattice theory, as in Section 5.3, and in field theory.

Notes to Chapter 9

This chapter is based on Negri and von Plato (2005). In von Plato (1995), the basic relations of constructive elementary geometry were treated.

The idea of an apartness relation in place of an equality relation appears first in Brouwer's works on the intuitionistic continuum from the early 1920s. Apartness was written as $a \# b$. We have used a uniform notation in which the intuitionistic notion is written with a slash over the classical one, as in $a \neq b$. Such apartness is in the discrete case equivalent to the negation of equality, or else it denotes the primitive notion of apartness. Brouwer also introduced a whole range of notions of apartness, often with very idiosyncratic notations or terminologies. For example, in his two papers (1927, 1950) a strict linear order $a < b$ was introduced, with the symmetrization $a < b \vee b < a$ corresponding to apartness, and the double negation $\neg\neg \, a < b$ giving 'the non-contradictory of the measurable natural order on the continuum'.

A formal treatment of apartness relations began with the formalization of elementary intuitionistic geometry in Heyting's doctoral dissertation of 1925; see Heyting (1927). In it, point and line equalities were replaced by corresponding apartnesses and the basic axiomatic properties studied. Heyting's little book (1956) presented intuitionistic axiomatizations of apartness and order relations. In the 1960s, such relations were used in different approaches to constructive analysis, as in Scott (1968), who established the notation we use. At the same time, there began a period when intuitionistic axiomatic systems were studied by the means of Kripke semantics, as in Smorynski (1973).

10 | Proof analysis in elementary geometry

The axiomatizations of plane projective and affine geometry include the axiom of non-collinearity, i.e., of the existence of at least three non-collinear points. It is shown that this axiom, when converted into a suitable rule, is conservative over the other rules in the following sense: if an atomic formula is derivable by all the rules from a given finite number of atomic formulas used as assumptions, it is derivable without the rule of non-collinearity. (Thus, a proper use of existential axioms requires existential conclusions.) By the subterm property for the rules with non-collinearity excluded, derivability by the rules of projective and affine geometry is decidable.

As an immediate application of the decision method, we conclude that any finite set of atomic formulas is consistent. As a second application, we prove the independence of the parallel postulate in affine geometry: a very short proof search is exhaustive but fails to give a derivation. Thus, we see, within the system of geometry, that no derivation can lead to the parallel postulate.

It should be noted that the solution to the decision problem for projective and affine geometries applies only to derivations by the geometric rules. When logical rules are applied, to conclude logically compound formulas, the decision problem is known to have, by a result announced first in Tarski (1949), a negative solution. Finally, it should be noted that the decision methods presented here are provably terminating algorithms of proof search. Earlier results in this direction have often given decidability in the weak sense of impossibility of undecidability and no upper bound.

10.1 Projective geometry

(a) **Basic relations, constructions, and axioms.** We have two types of basic objects, points and lines. 'Given' points are denoted by $a, b, c, \ldots$ and lines by $l, m, n, \ldots$ In addition, certain rules contain eigenvariables $x, y, z, \ldots$ The basic relations of projective geometry are as follows:

$a = b$, a and b are equal points,
$l = m$, l and m are equal lines,
$a \in l$,   point a is incident with line l.

Next we have two constructions:

$ln(a, b)$,   the connecting line of points a, b,
$pt(l, m)$,   the intersection point of lines l, m.

The geometric axioms have been presented in Chapter 3 and are not repeated here.

(b) Multiple-conclusion rules.

The geometric axioms will be converted into rules that are a multiple-conclusion generalization of those of natural deduction. The reason is that in such rules, each conclusion is the premiss of a successive rule or an endformula in a derivation. Permutation of rules is much more manageable than in sequent calculus in which a formula can remain inactive. A sequent calculus formulation is also possible.

Given an axiom of the general form $P_1 \& \ldots \& P_m \supset Q_1 \lor \ldots \lor Q_n$, the corresponding rule is

$$\frac{P_1 \ \ldots \ P_m}{Q_1 \ \ldots \ Q_n} \tag{1}$$

Here the atoms $P_1, \ldots, P_m$ are the **premisses** and the atoms $Q_1, \ldots, Q_n$ the **conclusions**, with $m, n \geqslant 0$. If $m = 0$, we have a **zero-premiss** rule, and if $n = 0$, a rule with an **empty conclusion**. The conclusions of a rule represent the several possible cases under the premisses of the rule. If rules are restricted to having just one conclusion, logical notation is needed to express cases.

The atoms of geometry contain free parameters and therefore the geometric rules are schematic. When values are given to these, a **rule instance** is obtained.

Definition 10.1. *A **derivation** is a finite combination of rule instances such that*

1. *Each formula occurrence is the premiss of at most one rule instance.*
2. *Each formula occurrence is the conclusion of at most one rule instance.*
3. *There is no sequence of rule instances such that, proceeding from the premisses of a rule instance upwards through rules in which the premisses are conclusions, one arrives in two ways at the same rule instance.*

We say that derivations that follow rule scheme (1) with conditions 1–3 of the definition are written in **natural deduction style**. The qualifications 'instance' and 'occurrence' are usually dropped when rule instances and formula occurrences in derivations are considered. Formulas in a derivation that are not conclusions of some rule are the (open) **assumptions** of the derivation. Formulas in a derivation that are not the premisses of some rule are the (open) **cases** of the derivation.

As a limiting case of zero rule instances, we have derivations of the form

$$P$$

in which the atom P is at the same time an assumption and a case.

Derivations as defined in 10.1 need not be representable in two dimensions. However, all those parts of derivations that we need to consider can be printed in two dimensions provided that we permit the reordering of premisses and conclusions. We refer to them as the first premiss (P_1 in the rule scheme), the second premiss, and so on, and similarly for the conclusions.

A **thread** in a derivation is a sequence of formulas $(P_1, \ldots, P_k)$ such that P_1 is an assumption, P_k a case, and P_i a premiss in a rule and P_{i+1} a conclusion in the same rule, with $1 \leqslant i < k$.

A derivation has a **loop** (or **cycle**) if it has a thread of the form

$$(P_1, \ldots, P, \ldots, P, \ldots, P_k)$$

The **branches** from a formula P in a derivation are formed by those sequences $(P_1, \ldots, P)$ for which there is a thread $(P_1, \ldots, P, \ldots, P_k)$ in the derivation, and the **roots** from P the sequences $(P, \ldots, P_k)$.

If the number of cases in each rule in a derivation is at most one, the derivation is in **tree form**. Dually, if the number of premisses is at most one, the derivation is in **root form**.

Given a derivation $\mathcal{D}$, its **upward subderivations** are the derivations obtained by deleting the roots of at least one conclusion formula P while maintaining P as a case. A **downward subderivation** is obtained by similarly deleting all branches of at least one premiss formula while maintaining the formula as an assumption.

The rules of logical inference do not interfere with derivations by mathematical rules of the form (1), because steps of logical inference can be permuted down relative to the mathematical rules. Therefore the question of the derivability of given atomic cases Δ from given atomic assumptions Γ concerns only the mathematical rules.

We can write multiple-conclusion rules in a sequent notation. The scheme (1) becomes

$$\frac{\Gamma_1 \to \Delta_1, P_1 \ldots \Gamma_m \to \Delta_m, P_m}{\Gamma_1, \ldots, \Gamma_m \to \Delta_1, \ldots, \Delta_m, Q_1, \ldots, Q_n}$$

The rule for composing derivations is in sequent calculus notation

$$\frac{\Gamma_1 \to \Delta_1, P \quad P, \Gamma_2 \to \Delta_2}{\Gamma_1, \Gamma_2 \to \Delta_1, \Delta_2} \; Comp$$

Lemma 10.2. Admissibility of composition. *The rule of composition is admissible in systems of rules that follow scheme* (1).

Proof. The proof is by induction on the number of steps of inference of the right premiss of rule *Comp*. We show that instances of *Comp* can be permuted up until they disappear.

The base case is the derivation $P \to P$, with the instance of *Comp*

$$\frac{\Gamma_1 \to \Delta_1, P \quad P \to P}{\Gamma_1 \to \Delta_1, P} \; Comp$$

Thus, the conclusion is equal to the left premiss and the instance of *Comp* can be deleted. In the inductive case, the last rule in the derivation of the right premiss of *Comp* has the form

$$\frac{P, \Gamma_{2_1} \to \Delta_{2_1}, P_1 \quad \cdots \quad \Gamma_{2_m} \to \Delta_{2_m}, P_m}{P, \Gamma_2 \to \Delta_{2_1}, \ldots, \Delta_{2_m}, Q_1, \ldots, Q_n} \; Rule$$

Here $\Gamma_{2_1}, \ldots, \Gamma_{2_m} \equiv \Gamma_2$, and $\Delta_{2_1}, \ldots, \Delta_{2_m}, Q_1, \ldots, Q_n \equiv \Delta_2$. We may assume the composition formula in the second premiss of *Comp* to come from the first premiss of *Rule*. By the inductive hypothesis, composition of the first premiss of *Comp* with $P, \Gamma_{2_1} \to \Delta_{2_1}, P_1$ is admissible, the conclusion being $\Gamma_1, \Gamma_{2_1} \to \Delta_1, \Delta_{2_1}, P_1$. Now *Rule* is applied with this sequent as the first premiss, and with the same conclusion as in the original instance of *Comp*. QED.

Consider a derivation of $\Gamma \to \Delta$ with a loop as in the thread:

$$(P_1, \ldots, P, \ldots, P, \ldots, P_k)$$

The loop can be eliminated as follows. The first occurrence of P of the thread determines an upward subderivation of a sequent $\Gamma_1 \to \Delta_1, P$, and the second a downward subderivation of a sequent $P, \Gamma_2 \to \Delta_2$. By the rule of composition, these can be put together into a derivation of $\Gamma_1, \Gamma_2 \to \Delta_1, \Delta_2$. Branches and roots between the two occurrences of P in the original derivation have been removed, so the result of composition gives a reduced multiset of assumptions and cases of the original derivation: a derivation that is sharper in the sense of having fewer cases derived from fewer assumptions.

In the natural deduction style of writing derivations, each conclusion is by definition a premiss in the next rule or a case of the whole derivation. In sequent calculus style, conclusions can instead remain inactive. It is, however, always possible to permute the order of application of sequent calculus rules so that a unique correspondence with a natural deduction style derivation is found. We shall therefore use the more readable natural deduction style in the proofs of the main Lemmas 10.3 and 10.5 below.

(c) **The rules of projective geometry.** The rules of projective geometry that correspond to axioms I–IV in Section 3.1(b) are given below. For axiom V, we use the co-geometric rule scheme of Chapter 9:

Table 10.1 The rules of projective geometry

I **Rules for equality relations**

$$\frac{}{a = a}\ Ref \qquad \frac{a = b}{b = a}\ Sym \qquad \frac{a = b \quad b = c}{a = c}\ Tr$$

$$\frac{}{l = l}\ Ref \qquad \frac{l = m}{m = l}\ Sym \qquad \frac{l = m \quad m = n}{l = n}\ Tr$$

II **Rules for incidence**

$$\frac{}{a \in ln(a, b)}\ ILn_1 \qquad \frac{}{b \in ln(a, b)}\ ILn_2$$

$$\frac{}{p\,t(l, m) \in l}\ IPt_1 \qquad \frac{}{p\,t(l, m) \in m}\ IPt_2$$

III **Uniqueness rule**

$$\frac{a \in l \quad a \in m \quad b \in l \quad b \in m}{a = b \qquad l = m}\ Uni$$

IV **Substitution rules**

$$\frac{a \in l \quad a = b}{b \in l}\ SPt \qquad \frac{a \in l \quad l = m}{a \in m}\ SLn$$

V **Rule of non-collinearity**

$$\frac{x = y}{}\ ET,1 \qquad \frac{z \in ln(x, y)}{}\ ET,1$$

Here $x = y$ and $z \in ln(x, y)$ are conclusions in a given derivation. The rule **closes** the possible cases $x = y$ and $z \in ln(x, y)$ in roots of a derivation, which is indicated by the empty conclusion. A numerical label is needed for showing which atoms go together in an application of the rule, similarly to the use of discharge labels next to rule symbols and above closed formulas in natural deduction. There can be any numbers $m \geqslant 0$, $n \geqslant 0$ of occurrences of $x = y$ and $z \in ln(x, y)$ that are closed in one rule instance (as long as there is at least one of either). In the rule, x, y, and z are the **eigenvariables**, assumed distinct and with no free occurrences of the eigenvariables in the remaining assumptions or cases of the derivation.

In words, rule ET states that if it can be derived that the cases under Γ include the case in which any two points x, y are equal, or (if not, then at least) the one in which any third point z is incident with the line $ln(x, y)$, then these two cases can be excluded.

Rule ET can be written in a 'local' style by the use of sequent calculus notation. The general case of arbitrary numbers $m \geqslant 0$, $n \geqslant 0$ of the two cases $x = y$ and $z \in ln(x, y)$ is indicated by $x = y^m$ and $z \in ln(x, y)^n$.

$$\frac{\Gamma \to \Delta, x = y^m, z \in ln(x, y)^n}{\Gamma \to \Delta} \; ET$$

The condition on the eigenvariables x, y, z is that they must not be free in the conclusion $\Gamma \to \Delta$ of the rule.

We show that rule ET together with the logical rules of classical sequent calculus makes the axiom of non-collinearity derivable:

$$\frac{\dfrac{\dfrac{x = y \to x = y}{\to x = y, \neg x = y} \, R\neg \quad \dfrac{z \in ln(x, y) \to z \in ln(x, y)}{\to z \in ln(x, y), \neg z \in ln(x, y)} \, R\neg}{\dfrac{\to x = y, z \in ln(x, y), \neg x = y \,\&\, \neg z \in ln(x, y)}{\dfrac{\to x = y, z \in ln(x, y), \exists x \exists y \exists z (\neg x = y \,\&\, \neg z \in ln(x, y))}{\to \exists x \exists y \exists z (\neg x = y \,\&\, \neg z \in ln(x, y))} \, ET} \, R3,R3,R3} \, R\&}$$

In the other direction, assuming the premiss of rule ET given, we obtain its conclusion from the axiom of non-collinearity by:

$$\frac{\to \exists x \exists y \exists z (\neg x = y \,\&\, \neg z \in ln(x, y)) \quad \dfrac{\dfrac{\dfrac{\Gamma \to \Delta, x = y^m, z \in ln(x, y)^n}{\neg x = y, \neg z \in ln(x, y), \Gamma \to \Delta} \, L\neg, L\neg}{\neg x = y \,\&\, \neg z \in ln(x, y), \Gamma \to \Delta} \, L\&}{\exists x \exists y \exists z (\neg x = y \,\&\, \neg z \in ln(x, y)), \Gamma \to \Delta} \, L3,L3,L3}{\Gamma \to \Delta} \; Comp$$

It follows that our rule calculus is equivalent to a standard axiomatic calculus for projective geometry.

The axioms of group II in Chapter 3 contained the degenerate cases of the constructions. These have been left out in the corresponding rules to make the presentation that follows simpler.

(d) The subterm property. As the main result, we shall prove a **subterm property** for loop-free derivations in the rule systems for plane projective and affine geometry. The latter proof is given in the next section.

Definition. A **new term** in a derivation of $\Gamma \to \Delta$ is one that is not a term in Γ, Δ. Terms in Γ, Δ are **known terms**. The **length** of a term is the number of geometric constructions in it.

We show how to remove possible new terms from derivations of atomic cases Δ from atomic assumptions Γ in projective geometry. Rule V is conservative over rules I–IV in such derivations, so the main lemma states the subterm property for derivations by rules I–IV. The proof is long with lots of cases. We therefore give first a summary description:

Outline of the proof of the main lemma. The rules of projective geometry have the remarkable property that each term in a conclusion is also a term in some premiss. Consider a new term of maximal length, say a line. We trace it to topformulas, and find that the only way it can appear in these is through the zero-premiss incidence rules, say in $a \in ln(a, b)$ or $b \in ln(a, b)$. In these, $ln(a, b)$ is a new term of maximal length, and no ground terms can be new terms of maximal length. Terms are removed from a derivation only through transitivity and substitution. In both, the new term $ln(a, b)$ occurs in an equality. The only way terms first appearing in an incidence can be found further down in an equality is through rule *Uni*. We transform the uppermost instances of *Uni* with the term $ln(a, b)$ in a conclusion so that in the next step one of the following cases appears:

A   The new term is removed by *Tr*.
B   The new term is removed by *SLn*.
C   The new term reappears in an incidence as a conclusion of rule *SLn*.

In case **A**, the other premiss of *Tr* is also an equation with $ln(a, b)$. If it is a conclusion of an uppermost instance of *Uni*, a proof transformation removes the new term. If it is not, we trace up the term $ln(a, b)$ until an uppermost *Uni* and find again one of **A**, **B**, or **C**. Case **B** leads either to the removal of $ln(a, b)$ or to case **A**. In case **C**, there must be a second instance of *Uni* concluding an equation with the term $ln(a, b)$. Permutations lead either to case **A** or to a second-to-uppermost *Uni* with $ln(a, b)$. Proof transformations are given that reduce this situation to the previously covered ones.

Lemma 10.3. *If the atomic cases Δ are derivable from the atomic assumptions Γ with rules I–IV of plane projective geometry, there is a derivation with no new terms.*

Proof. The proof is divided into parts indicated by boldface numbers.

1. First occurrences of new terms. Consider a new term in a loop-free derivation, say a line *l*. The following condition can be put:

Condition 1. The term l is a term of maximum length among all new terms in the derivation and the first in the lexicographical ordering of such terms.

Consider a downmost occurrence of a new term. The atom in which it occurs determines an upward subderivation in which we trace up, step by step, atoms with the new term, until we arrive at topformula atoms. For l, these atoms occur in instances of *Ref*, *IPt*, or *ILn*. The first is excluded because all applicable rules with $l = l$ as one premiss give loops. With *IPt* we have, say, $pt(l, m) \in l$, but then $pt(l, m)$ is a new term longer than l, against assumption 1. Therefore only ILn_1 and ILn_2 are possible and we have the result that l is identical to a line $ln(a, b)$ for some points a, b, and the possible topformulas are

$$\frac{}{a \in ln(a, b)}\ ILn_1 \qquad \frac{}{b \in ln(a, b)}\ ILn_2 \qquad .$$

2. Rules that remove new terms. The rules that can remove the new term $ln(a, b)$ are *Tr* and *SLn*. In both, $ln(a, b)$ occurs as a term in an equation in a premiss. The only rule that can introduce such an equation in the derivation is *Uni*, say, one of

$$\frac{c \in ln(a, b) \quad c \in m \quad d \in ln(a, b) \quad d \in m}{c = d \quad ln(a, b) = m}\ Uni$$

$$\frac{c \in m \quad c \in ln(a, b) \quad d \in m \quad d \in ln(a, b)}{c = d \quad m = ln(a, b)}\ Uni \tag{1}$$

If m is identical to $ln(a, b)$, the conclusion contains the reflexivity atom $ln(a, b) = ln(a, b)$. When this atom is a premiss in *Sym*, *Tr*, or *SLn*, a loop is found. Therefore the second and fourth (resp. first and third) premisses do not contain the term $ln(a, b)$. If (1) is an uppermost instance of *Uni* with the term $ln(a, b)$ in a conclusion, $ln(a, b)$ does not occur in an equation in any upward subderivation starting from these premisses.

We prove the lemma by showing the following:

1. *One instance of Uni.* If the maximal number of instances of *Uni* with $ln(a, b)$ in a conclusion is 1 in the threads of a derivation, these instances can be converted so that no equation with the term $ln(a, b)$ is concluded. Therefore there cannot be any new term $ln(a, b)$ left.

2. *Reduction of the number of Uni in the threads.* If the maximal number of instances of *Uni* with $ln(a, b)$ in a conclusion is more than 1 in the threads of a derivation, it can be reduced.

Base case *1* and inductive case *2* mix in rather intricate ways in the proof.

3. The form of an uppermost uniqueness. Assume that (1) is an uppermost instance of *Uni* with the term $ln(a, b)$ in a conclusion, and consider its first premiss $c \in ln(a, b)$. The only rule other than *ILn* that can have this premiss as a conclusion is *SPt*, and the same for the first premiss of this instance of *SPt* until an instance of *ILn* is found. We may assume this to be the instance $a \in ln(a, b)$:

$$\frac{\dfrac{\rule{2cm}{0.4pt}}{a \in ln(a, b)}{}^{ILn_1} \quad a = a_1}{\dfrac{a_1 \in ln(a, b)}{} {}^{SPt} \quad \dfrac{\quad a_1 = a_2}{}{}^{SPt}}$$

$$\frac{a_m \in ln(a, b) \qquad a_m = c}{c \in ln(a, b)}{}^{SPt} \tag{2}$$

We can derive $a = c$ by *Tr* from the right premisses in (2):

$$\frac{a = a_1 \quad a_1 = a_2}{a = a_2}{}^{Tr}$$
$$\vdots$$
$$a = c \tag{3}$$

For the third premiss $d \in ln(a, b)$, we have by analogous arguments

$$b = d \tag{4}$$

In (4) it is assumed that a left premiss $b_n \in ln(a, b)$ of *SPt*, analogous to the premiss $a_m \in ln(a, b)$ in (2), led to a topformula instance $b \in ln(a, b)$. In the contrary case of $a \in ln(a, b)$, (4) would be the conclusion $a = d$. From this, together with (3), the case $c = d$ of (1) could then be derived without instances of *Uni*.

We now transform (1) by the use of (2)–(4) into

$$\frac{\dfrac{\dfrac{\rule{2cm}{0.4pt}}{a \in ln(a,b)}{}^{ILn_1} \quad a \overset{\vdots}{=} c}{c \in ln(a,b)}{}^{SPt} \quad c \in m \quad \dfrac{\dfrac{\rule{2cm}{0.4pt}}{b \in ln(a,b)}{}^{ILn_2} \quad b \overset{\vdots}{=} d}{d \in ln(a,b)}{}^{SPt} \quad d \in m}{c = d \quad ln(a, b) = m}{}^{Uni} \tag{5}$$

and similarly with the conclusion $m = ln(a, b)$.

4. Rules that have as a conclusion an equation with the new term. The term $ln(a, b)$ appears first in an equation only as a conclusion of *Uni* and remains in an equation in a conclusion as long as such an equation is a premiss in *Sym* or a premiss in *Tr* with $ln(a, b)$ not the middle term. We show that such instances of *Sym* and *Tr* can be deleted or permuted above *Uni* (5).

Assume then that there is a root down from $ln(a, b) = m$ in (5) with a sequence of instances of *Sym* and *Tr* until a step of *Tr* in which $ln(a, b)$ is a middle term or until a step of *SLn* with $ln(a, b)$ in the second premiss. If this branch has two consecutive instances of *Sym*, delete both. If it has two consecutive instances of *Tr* we have, say

$$\frac{\dfrac{ln(a, b) = m \quad m = n}{ln(a, b) = n} \, Tr \quad n = l}{ln(a, b) = l} \, Tr \tag{6}$$

with l, m, n distinct from $ln(a, b)$. This is transformed into

$$\frac{ln(a, b) = m \quad \dfrac{m = n \quad n = l}{m = l} \, Tr}{ln(a, b) = l} \, Tr \tag{7}$$

In the end, we have an alternating sequence of single instances of *Sym* and *Tr*. Assume that there are at least two instances of *Sym*, with the part of derivation

$$\frac{n = m \quad \dfrac{ln(a, b) = m}{m = ln(a, b)} \, Sym}{\dfrac{n = ln(a, b)}{ln(a, b) = n} \, Sym} \, Tr \tag{8}$$

Transform this into

$$\frac{ln(a, b) = m \quad \dfrac{n = m}{m = n} \, Sym}{ln(a, b) = n} \, Tr \tag{9}$$

If the conclusion of (9) is a first premiss in *Tr*, a transformation as in (7) is made to remove one *Tr* with line $ln(a, b)$.

In the end, there is at most one instance of *Sym* and *Tr*, say

$$\frac{l = m \quad \dfrac{\dfrac{c \in ln(a, b) \quad c \in m \quad d \in ln(a, b) \quad d \in m}{ln(a, b) = m} \, Uni}{m = ln(a, b)} \, Sym}{l = ln(a, b)} \, Tr \tag{10}$$

This is transformed into

$$\frac{\dfrac{c \in m \quad \dfrac{l = m}{m = l} \, Sym}{c \in l} \, SLn \quad c \in ln(a, b) \quad \dfrac{d \in m \quad \dfrac{l = m}{m = l} \, Sym}{d \in l} \, SLn \quad d \in ln(a, b)}{l = ln(a, b)} \, Uni \tag{11}$$

If the instances of *Sym* and *Tr* are in the other order as compared with (10), the transformation is the same as (11).

If there is just one instance of *Sym* that lets us derive $m = ln(a, b)$ from $ln(a, b) = m$ and no *Tr*, we interchange the first and second premiss, and the third and fourth premiss, respectively, of *Uni* (5), to obtain the conclusion without *Sym*. If there is just one instance of *Tr* with $ln(a, b)$ in the conclusion, we have one of

$$\frac{ln(a, b) = m \quad m = n}{ln(a, b) = n} \, Tr \qquad \frac{l = m \quad m = ln(a, b)}{l = ln(a, b)} \, Tr \qquad (12)$$

The second is transformed as in (11), and the first analogously.

By the above permutations, we are left with three possible ways in which the term $ln(a, b)$ in the conclusion of (5) can appear as a premiss in a successive rule:

A $ln(a, b)$ is a middle term in a premiss in *Tr*.
B $ln(a, b)$ is the left term in a second premiss of *SLn*.
C $ln(a, b)$ is the right term in a second premiss of *SLn*.

These cases correspond to those in the outline of the proof. Cases **A** and **B** remove the new term. These are treated in **5** below and case **C** in **6**. We now put the

Condition 2. Uppermost instances of *Uni* with $ln(a, b)$ in a conclusion have been transformed as in **3** and **4**.

5. Removal of new terms. We assume first that there is at most one instance of *Uni* with $ln(a, b)$ in a conclusion along threads. The term $ln(a, b)$ in the conclusion of (5) is removed by *Tr* or *SLn* and we have two subcases:

5.1. If the rule is *Tr*, we have

$$\frac{l = ln(a, b) \quad ln(a, b) = m}{l = m} \, Tr \qquad (13)$$

By assumption, also the first premiss is a conclusion of an uppermost instance of *Uni* and we have, say,

$$\frac{c' \in l \quad c' \in ln(a, b) \quad d' \in l \quad d' \in ln(a, b)}{l = ln(a, b)} \, Uni \qquad (14)$$

(The first conclusion $c' = d'$ of (14) can be left unwritten here and later.) We conclude as for (3) and (4) that

$$a = c' \qquad (15)$$

and

$$b = d' \tag{16}$$

are derivable. The upward subderivations from these equations do not contain the term $ln(a, b)$ in an equation. The step of Tr that removes $ln(a, b)$ now has a first premiss given by (14) and a second premiss given by (5). It is transformed, with some obvious instances of Sym left unwritten, into

$$\cfrac{c' \in l \quad \cfrac{\cfrac{c' = a \quad a = c}{c' = c}\,Tr}{c = l}\,SPt}{c \in l} \qquad \cfrac{\cfrac{c \in m \qquad \cfrac{d' \in l \quad \cfrac{\cfrac{d' = b \quad b = d}{d' = d}\,Tr}{d = l}\,SPt}{d \in l}}{l = m} \qquad \cfrac{}{d \in m}\,Uni}{} \tag{17}$$

The instances of Uni concluding equations with the term $ln(a, b)$ have been removed.

5.2. If the term $ln(a, b)$ is removed by SLn, we have

$$\cfrac{e \in ln(a, b) \quad ln(a, b) = m}{e \in m}\,SLn \tag{18}$$

We consider the derivation of the first premiss. Possible rules are ILn, SPt, and SLn and we have three sub-subcases:

5.2.1. The first premiss of (18) is an instance of ILn that we may assume without loss of generality to be $a \in ln(a, b)$. Then e is identical to a and the derivation is:

$$\cfrac{\cfrac{}{a \in ln(a, b)}\,ILn_1 \quad ln(a, b) = m}{a \in m}\,SLn \tag{19}$$

It is converted, by the use of the second premiss of (5) and by (3), into

$$\cfrac{c \in m \quad c = a}{a \in m}\,SPt \tag{20}$$

The instance of Uni with the term $ln(a, b)$ in the conclusion has been removed.

5.2.2. The first premiss of (18) is concluded by rule SPt and we have

$$\cfrac{\cfrac{f \in ln(a, b) \quad f = e}{e \in ln(a, b)}\,SPt \quad ln(a, b) = m}{e \in m}\,SLn \tag{21}$$

Rule *SLn* is permuted up:

$$\frac{\dfrac{f \in ln(a, b) \quad ln(a, b) = m}{f \in m} \; SLn \qquad f = e}{e \in m} \; SPt \tag{22}$$

This case leads to the removal on *Uni* as in **5.2.1**.

5.2.3. The first premiss of (18) is concluded by rule *SLn* and we have

$$\frac{\dfrac{e \in l \quad l = ln(a, b)}{e \in ln(a, b)} \; SLn \qquad ln(a, b) = m}{e \in m} \; SLn \tag{23}$$

The derivation is transformed into

$$\frac{e \in l \quad \dfrac{l = ln(a, b) \quad ln(a, b) = m}{l = m} \; Tr}{e \in m} \; SLn \tag{24}$$

This case leads to a step of transitivity removing $ln(a, b)$, covered by case **5.1**.

Assume next that there is along threads more than one instance of *Uni* with the term $ln(a, b)$. If the first premiss in (13) is not a conclusion of an uppermost instance of *Uni* with the term $ln(a, b)$, we trace up $ln(a, b)$ until an uppermost instance. Now one of the cases **A–C** at the end of **4** applies and we continue the proof analysis as in **5**. If in repeating this process the term $ln(a, b)$ is never a right term in the second premiss of *SLn* (case **C**), occurrences of $ln(a, b)$ are removed in the same way as in **5.1** and **5.2.1**. Otherwise case **C** is met:

6. New term in an incidence. We are left with the case in which the term $ln(a, b)$ in an equation in (5) is a premiss in *SLn* after which it appears in an incidence:

$$\frac{e \in m \quad \dfrac{c \in m \quad c \in ln(a, b) \quad d \in m \quad d \in ln(a, b)}{m = ln(a, b)} \; Uni}{e \in ln(a, b)} \; SLn \tag{25}$$

The conclusion can be a premiss in *SPt*, *SLn*, or *Uni* and we have three subcases:

6.1. If $e \in ln(a, b)$ is a premiss in *SPt*, the other premiss is some $e = f$ and *SPt* permutes to the first premiss of *SLn*:

$$\frac{\dfrac{e \in m \quad m = ln(a, b)}{e \in ln(a, b)} \; SLn \qquad e = f}{f \in ln(a, b)} \; SPt \quad \rightsquigarrow \quad \frac{\dfrac{e \in m \quad e = f}{f \in m} \; SPt \qquad m = ln(a, b)}{f \in ln(a, b)} \; SLn \tag{26}$$

Eventually an instance of *SLn* or *Uni* is found:

6.2. If $e \in ln(a, b)$ is a premiss in SLn, the other premiss is some $ln(a, b) = n$ and the derivation and its permutation are

$$\dfrac{\dfrac{e \in m \quad m = ln(a, b)}{e \in ln(a, b)}SLn \quad ln(a, b) = n}{e \in n}SLn \rightsquigarrow \dfrac{e \in m \quad \dfrac{m = ln(a, b) \quad ln(a, b) = n}{m = n}Tr}{e \in n}SLn \tag{27}$$

If there is just one Uni with $ln(a, b)$ along threads, case **5.1** applies. If not, we trace up from $ln(a, b) = n$ to an uppermost Uni with $ln(a, b)$.

6.3. Eventually the transformations in **6.1** and **6.2** lead to the third and final subcase of **6** in which $e \in ln(a, b)$ is a premiss in Uni. The conclusion has the term $ln(a, b)$ and the relevant part of the derivation is:

$$\dfrac{e \in m \quad \dfrac{\dfrac{c \in m \quad c \in ln(a, b) \quad d \in m \quad d \in ln(a, b)}{m = ln(a, b)}Uni}{e \in ln(a, b)}SLn \quad \dfrac{e \in n \quad f \in ln(a, b)}{f \in n}Rule}{ln(a, b) = n}Uni \tag{28}$$

$Rule$ can be an instance of ILn, SPt, or SLn and we have three sub-subcases:

6.3.1. If $Rule$ is ILn with, say, f identical to a, (28) is transformed into

$$\dfrac{c \in ln(a, b) \quad \dfrac{\dfrac{a \in n \quad c = a}{c \in n}SPt \quad d \in ln(a, b)}{ln(a, b) = n}}{\dfrac{d \in m \quad \dfrac{\dfrac{c \in m \quad c = a}{a \in m}SPt \quad \dfrac{a \in n \quad e \in m \quad e \in n}{m = n}SLn}{d \in n}Uni}{}}Uni \tag{29}$$

The number of instances of Uni along threads with $ln(a, b)$ in the conclusion has been reduced.

6.3.2. If $Rule$ is SPt with a second premiss $g = f$, the order of the premisses of the uppermost Uni in (28) is changed to allow us to conclude $ln(a, b) = m$ (twice) and (28) is transformed into

$$\dfrac{ln(a, b) = m \quad \dfrac{e \in m \quad e \in n}{ln(a, b) = n}}{\dfrac{\dfrac{g \in ln(a, b) \quad g = f}{f \in ln(a, b)}SPt \quad \dfrac{ln(a, b) = m}{f \in m}SLn}{\dfrac{m = n}{}Tr}}{f \in n}Uni \tag{30}$$

The upward subderivation from $f \in m$ is covered by case **5.2.** Note that the last step is at this point not yet transformed as in **3** and **4**, but only after the term $ln(a, b)$ has been removed from the subderivation.

6.3.3. If *Rule* is *SLn*, it has a second premiss $l = ln(a, b)$ and the derivation is

$$\frac{\dfrac{e \in m \quad m = ln(a, b)}{e \in ln(a, b)} \, SLn \quad \dfrac{\dfrac{f \in l \quad l = ln(a, b)}{f \in ln(a, b)} \, SLn \quad f \in n}{ \quad } \, Uni}{ln(a, b) = n} \tag{31}$$

There are two threads that contain at least two instances of *Uni* with $ln(a, b)$ in a conclusion. The transformed derivation is

$$\frac{ln(a, b) = l \quad \dfrac{e \in m \quad \dfrac{\dfrac{l = ln(a, b) \quad ln(a, b) = m}{l = m} \, Tr}{e \in l} \, SLn \quad \dfrac{e = n \quad f \in l \quad f \in n}{l = n} \, Uni}{ln(a, b) = n}}{ } \, Tr \tag{32}$$

If $l = ln(a, b)$ is a conclusion of an uppermost instance of *Uni*, the subderivation down to $l = m$ is transformed as in (17) with both of the uppermost instances of *Uni* removed from the derivation.

If $l = ln(a, b)$ is not a conclusion of an uppermost *Uni*, we trace up $ln(a, b)$ until an uppermost instance is found. All cases that can arise are now covered.

By the above process, all occurrences of $ln(a, b)$ are removed. If a longest new term is a point, it is removed in a process dual to the above. By repeating the removal all new terms, in descending order of length, are removed.   QED.

We show next that the existence of non-collinear points, rule V, is conservative over rules I–IV for derivations that contain only atomic formulas:

Lemma 10.4. *If the atomic cases Δ are derivable from the atomic assumptions Γ with rules I–V of plane projective geometry, they are already derivable with rules I–IV.*

Proof. We write the proof using sequent calculus notation. Assume that there is a loop-free derivation of $\Gamma \to \Delta$ with instances of rule *ET*, and consider a first such instance, with $m \geqslant 0$, $n \geqslant 0$ copies of the two formulas closed by the rule:

$$\frac{\Gamma \to \Delta', x = y^m, z \in ln(x, y)^n}{\Gamma \to \Delta'} \, ET$$

The premiss is derived by rules I–IV, so Lemma 10.3 applies to it. If a first occurrence of the eigenvariables x, y, z is found in an initial sequent, there are eigenvariables in the antecedent, against the variable restrictions. Therefore such occurrences are possible only in instances of *Ref* or *ILn*. The former is excluded, because all possible rules with a reflexivity atom give a loop. For rules *ILn*, the only known line term with eigenvariables in the derivation of the premiss of rule *ET* is $ln(x, y)$ and first occurrences of eigenvariables in *ILn* must therefore be of the form

$$\frac{}{\rightarrow x \in ln(x, y)}\,ILn_1 \qquad \frac{}{\rightarrow y \in ln(x, y)}\,ILn_2$$

There is no way in which the term z could occur in a topsequent and therefore there is no atom $z \in ln(x, y)$ in the derivation, so $n = 0$. This leaves the derivable sequent $\Gamma \rightarrow \Delta', x = y^m$, with x, y not terms in Γ, Δ'. Tracing x and y up to topsequents, *Ref* is excluded as before. Connecting lines containing x or y and occurring in Γ, Δ' would violate the variable restrictions. Thus, such lines are new terms and there is no instance of *ILn* containing x or y, so also $m = 0$. The conclusion of the rule is identical to its premiss, so the instance of rule *ET* can be removed.　　QED.

Lemmas 10.3 and 10.4 give us immediately the following Theorem:

Theorem 10.5. *If the atomic cases Δ are derivable from the atomic assumptions Γ with rules I–V of plane projective geometry, there is a derivation in which all terms are terms in Γ, Δ.*

Given a problem of derivability of atomic cases Δ from atomic assumptions Γ in projective geometry, proof search can be limited to the terms known from Γ, Δ. There is a bounded number of distinct atomic formulas with terms from Γ, Δ. Therefore there is also a bounded number of loop-free derivations, and derivability is decidable. Derivations are in tree form except when there is an instance of rule *Uni*. For the fragment without *Uni*, derivability can be decided, by well known results, by a polynomial-time algorithm, but not so for the full system.

　　To finish this section, we shall indicate how the proof of Lemma 10.3 could in principle, if sufficiently broad pages were available, be reproduced in a sequent calculus formulation of the geometrical rules. The essential property of the multiple-conclusion natural deduction formulation was that each conclusion of a rule was at once the premiss in a successive rule. A derivation is translated into a sequent notation by collecting all the open assumptions into a multiset Γ and all the cases into another multiset Δ. In the proof of Lemma 10.3, instances of rules I–IV are such that only one formula is active in a premiss. Therefore, since there are no eigenvariables that could block a permutation, the order of rules can be so transformed that a given atomic formula P active in a given rule was the principal formula in a preceding rule. Next, we considered an uppermost instance of rule *Uni*, the only rule with more than one principal formula. Therefore there are above it only rules with the above-mentioned property of permutability. It is now a routine matter to check that the proof transformations (28) to (32), with more than one instance of *Uni*, can be carried through when they are written in sequent notation. Overall, the only thing that happens when the proof is carried through in sequent notation is that sequent arrows and commas and contexts are added, but the proof transformations themselves remain as they are.

10.2 Affine geometry

To obtain an axiom system for affine geometry, the following additions and modifications are made to the projective axiomatization of Section 10.1:

There is one additional relation and construction:

(4.1) $l \parallel m$, l and m are parallel lines.

(4.2) $par(l, a)$,   the parallel to line l through point a.

The additional affine axioms are

I General axioms for parallelism

$l \parallel l$, $l \parallel m \supset m \parallel l$, $l \parallel m \& m \parallel n \supset l \parallel n$.

II Affine axioms of incidence and parallelism

$a \in par(l, a)$, $par(l, a) \parallel l$.

III Affine uniqueness axiom

$a \in l \& a \in m \& l \parallel m \supset l = m$.

IV Affine substitution axiom

$l \parallel m \& m = n \supset l \parallel n$.

The rules to be added to projective geometry are:

Table 10.2 The rules of affine geometry

I Rules for the parallelism relation

$$\frac{}{l \parallel l}\,Ref \qquad \frac{l \parallel m}{m \parallel l}\,Sym \qquad \frac{l \parallel m \quad m \parallel n}{l \parallel n}\,Tr$$

II Rules for incidence and parallelism

$$\frac{}{a \in par(l,\,a)}\,IA \qquad \frac{}{par(l,\,a) \parallel l}\,Par$$

III Uniqueness of parallels

$$\frac{a \in l \quad a \in m \quad l \parallel m}{l = m}\,Unipar$$

IV Substitution rule

$$\frac{l \parallel m \quad m = n}{l \parallel n}\,SA$$

It will be useful to distinguish between transitivity for line equality and parallelism by writing *TrLn* and *TrPar*.

We shall next prove the subterm property for loop-free derivations in affine geometry.

Lemma 10.5. *If the atomic cases Δ are derivable from the atomic assumptions Γ with rules I–IV of plane affine geometry, there is a derivation with no new terms.*

Proof. The proof is an extension of the proof for projective geometry in Lemma 10.3 and we consider only the new cases. The numbering of parts of the proof is as in lemma 10.3 and the numbering of derivations continues that of 10.3.

1. First occurrences of new terms. Consider a line $par(l,\,a)$ that is a new term of maximal length in a loop-free derivation. First occurrences of the term are in

$$\frac{}{a \in par(l,\,a)}\,IA \qquad \frac{}{par(l,\,a) \parallel l}\,Par$$

2. Rules that remove new terms. The term $par(l,\,a)$ can be removed by rules *TrLn, TrPar, SLn*, and *SA*. We consider first the case in which $par(l,\,a)$ does not occur in an equality in the derivation. Only *TrPar* can remove it:

$$\frac{m \parallel par(l,\,a) \quad par(l,\,a) \parallel n}{m \parallel n}\,TrPar \tag{33}$$

If either premiss is derived by *TrPar*, the latter is permuted above it as in (6) and (7). If the right premiss has been derived by *SA*, *TrPar* is again permuted above. If both premisses are derived by *Sym*, we have

$$\dfrac{\dfrac{par(l,\,a)\,\|\,m}{m\,\|\,par(l,\,a)}\,Sym \qquad \dfrac{n\,\|\,par(l,\,a)}{par(l,\,a)\,\|\,n}\,Sym}{m\,\|\,n}\,TrPar \tag{34}$$

A single step of *TrPar* followed by *Sym* gives the conclusion from the two premisses of the two *Sym*. In the end, we arrive at the derivation

$$\dfrac{\dfrac{\overline{par(l,\,a)\,\|\,l}\,{}^{Par}}{l\,\|\,par(l,\,a)}\,Sym \qquad \dfrac{}{par(l,\,a)\,\|\,l}\,Par}{l\,\|\,l}\,TrPar \tag{35}$$

The conclusion follows as an instance of *Ref*, with the part of derivation containing the new term deleted.

We can now assume that the term $par(l,\,a)$ appears in an equality in the derivation. Rules that can introduce $par(l,\,a)$ in an equality are

$$\dfrac{c \in par(l,\,a) \quad c \in m \quad d \in par(l,\,a) \quad d \in m}{c = d \quad par(l,\,a) = m}\,Uni \tag{36}$$

$$\dfrac{c \in par(l\,a) \quad c \in m \quad par(l\,a)\,\|\,m}{par(l\,a) = m}\,Unipar \tag{37}$$

and similarly with $par(l,\,a)$ as the right term of the equality.

3. The form of an uppermost uniqueness. Let (37) be an uppermost uniqueness rule with $par(l,\,a)$ in the conclusion. The first premiss $c \in par(l,\,a)$ has been derived, as in (2), by point substitutions, and the first occurrence of the new term is in $a \in par(l,\,a)$. The derivation is transformed, as in (3), into

$$\dfrac{\dfrac{}{a \in par(l,\,a)}\,IA \qquad \vdots \atop a = c}{c \in par(l,\,a)}\,SPt \tag{38}$$

In the derivation of the third premiss $par(l,\,a)\,\|\,m$ of (37), only rules *Par*, *Sym*, *TrPar*, and *SA* can appear, with $par(l,\,a)$ not in the second premiss of the last one. The derivation transforms into one of

$$\dfrac{\dfrac{}{par(l,\,a)\,\|\,l}\,Par \qquad \vdots \atop l\,\|\,m}{par(l,\,a)\,\|\,m}\,TrPar \tag{39}$$

or

$$\cfrac{\cfrac{}{par(l,\,a)\,\|\,l}{}^{Par} \qquad \cfrac{\vdots}{l=m}}{par(l,\,a)\,\|\,m}\,{}_{SA}$$

(40)

If rule *Uni* is, as in (36), the uppermost uniqueness rule, consider the premisses $c \in par(l,\,a)$ and $d \in par(l,\,a)$. As in (38), we conclude $a = c$ and $a = d$. Therefore $c = d$, the first conclusion of (36), follows, and the instance of *Uni* can be deleted. We can now assume that first occurrences of $par(l,\,a)$ in equations are in conclusions of rule *Unipar*.

4. Rules that have as a conclusion an equation with the new term. The new term introduced into an equation by (37) remains in an equation if the equation is a premiss of *Sym* or a premiss of *TrLn* with $par(l,\,a)$ not the middle term. The analysis proceeds as in (6)–(9), with the result that there is at most one *Sym* and *TrLn* following *Unipar*, say

$$\cfrac{k=m \qquad \cfrac{\cfrac{c \in par(l,\,a) \quad c \in m \quad par(l,\,a)\,\|\,m}{par(l,\,a)=m}\,{}_{Unipar}}{\cfrac{m=par(l,\,a)}{}}\,{}_{Sym}}{k=par(l,\,a)}\,{}_{TrLn}$$

(41)

This is transformed into

$$\cfrac{c \in par(l,\,a) \qquad \cfrac{c \in m \quad \cfrac{k=m}{m=k}\,{}_{Sym}}{c \in k}\,{}_{SLn} \qquad \cfrac{\cfrac{par(l,\,a)\,\|\,m \quad \cfrac{\cfrac{k=m}{m=k}\,{}_{Sym}}{\;}\,{}_{SA}}{par(l,\,a)\,\|\,k} \quad}{k\,\|\,par(l,\,a)}\,{}_{Sym}\,{}_{Unipar}}{k=par(l,\,a)}$$

(42)

If the instances of *Sym* and *TrLn* are in the other order, the transformation is similar.

We now have $k = par(l,\,a)$ (or $par(l,\,a) = k$) as a conclusion of an uppermost *Unipar*. The step following it is one of

A   *TrLn* with $par(l,\,a)$ the middle term.
B   *SLn* with $par(l,\,a)$ the removed term.
C   *SLn* with $par(l,\,a)$ the right term in the second premiss.
D   *SA* with $par(l,\,a)$ a term in the second premiss.

5. Removal of new terms. We consider cases **A** and **B** that remove the new term.

In case **A** we have a step of *TrLn* with the premisses $k = par(l, a)$ and $par(l, a) = k'$. By the transformations as in **4**, both premisses become conclusions of *Unipar*, as in

$$\dfrac{\dfrac{c \in k \quad c \in par(l, a) \quad k \parallel par(l, a)}{k = par(l, a)} \; Unipar \qquad \dfrac{c' \in par(l, a) \quad c' \in k' \quad par(l, a) \parallel k'}{par(l, a) = k'} \; Unipar}{k = k'} \; TrLn \qquad (43)$$

By (38), $a = c$ and $a = c'$, so $c = c'$. Derivation (43) is transformed into

$$\dfrac{c \in k \quad \dfrac{\dfrac{c' \in k' \quad c' = c}{c \in k'} \; SPt \quad \dfrac{k \parallel par(l, a) \quad par(l, a) \parallel k'}{k \parallel k'} \; TrPar}{k = k'} \; Unipar}{} \qquad (44)$$

The instances of *Unipar* with $par(l, a)$ in the conclusion have been removed.

In case **B**, the term $par(l, a)$ is removed by *SLn* and the conclusion of *Unipar* is of the form $par(l, a) = k$, so we have the steps

$$\dfrac{d \in par(l, a) \quad \dfrac{\dfrac{c \in par(l, a) \quad c \in k \quad par(l, a) \parallel k}{par(l, a) = k} \; Unipar}{d \in k} \; SLn}{} \qquad (45)$$

If the first premiss is an instance of *IA*, then d is identical to a. We have, as in (38), $a = c$, so the premiss $c \in k$ gives $a \in k$ without *Unipar*.

The other cases of derivation of the first premiss are treated similarly to the proof of Lemma 10.3, cases 5.2.2 and 5.2.3.

We have now covered cases **A** and **B**.

6. Now consider case **C** in which $par(l, a)$ is a right term in the second premiss of *SLn*. We have the steps

$$\dfrac{e \in k \quad \dfrac{\dfrac{c \in k \quad c \in par(l, a) \quad k \parallel par(l, a)}{k = par(l, a)} \; Unipar}{e \in par(l, a)} \; SLn}{} \qquad (46)$$

As for the projective case, consider rules in which the conclusion is a premiss:

6.1. $e \in par(l, a)$ is a premiss in *SPt*. The case is treated as the one for projective geometry.

6.2. $e \in par(l, a)$ is a premiss in *SLn*. The case is treated as the one for projective geometry.

6.3. $e \in par(l, a)$ is a premiss in *Unipar*. We have the steps

$$\dfrac{e \in k \quad \dfrac{\dfrac{c \in k \quad c \in par(l, a) \quad k \parallel par(l, a)}{k = par(l, a)} \; Unipar}{e \in par(l, a)} \; SLn \qquad \dfrac{e \in n \quad par(l, a) \parallel n}{par(l, a) = n} \; Unipar}{} \qquad (47)$$

We may assume (47) to be an uppermost case of its kind. Therefore $par(l, a)$ does not appear in an equation in the derivation of the third premiss of either *Unipar*. Analysis of the derivations of these premisses, as in (39) and (40) of **3**, shows that one of $l = k$ and $l \parallel k$ is derivable for the upper *Unipar*, and one of $l = n$ and $l \parallel n$ for the lower one. We have then altogether four cases:

1. $l = k$ and $l = n$. The conclusion $par(l, a) = n$ follows from the conclusion of the upper *Unipar* by steps of *TrLn*, so the case reduces to **A**.
2. $l = k$ and $l \parallel n$. Now $k \parallel n$ follows, so $e \in k, e \in n$ give by rule *Unipar* the conclusion $k = n$. Then $par(l, a) = n$ follows as in 1.
3. $l \parallel k$ and $l = n$. This goes through as 2.
4. $l \parallel k$ and $l \parallel n$. Now $k \parallel n$ and this goes through as 2.

Each of these cases reduces to case **A**.

6.4. $e \in par(l, a)$ is a premiss in *Uni*. The steps of derivation are

$$
\cfrac{e \in k' \quad \cfrac{e \in k \quad \cfrac{\cfrac{c \in k \quad c \in par(l, a) \quad k \parallel par(l, a)}{k = par(l, a)}\ Unipar}{e \in par(l, a)}\ SLn}{par(l, a) = n} \quad \cfrac{f \in k' \quad f \in par(l, a)}{}\ Uni} \tag{48}
$$

As in (38), $c \in par(l, a)$ gives $a = c$ and $f \in par(l, a)$ likewise $a = f$. Therefore $f \in k$ follows from $c \in k$, and we have the transformed steps

$$
\cfrac{e \in k' \quad \cfrac{e \in k \quad f \in k' \quad f \in k}{k' = k}\ Uni}{e = f} \qquad \cfrac{c \in k \quad \cfrac{c \in par(l, a) \quad k \parallel par(l, a)}{k = par(l, a)}\ Unipar}{k' = par(l, a)}\ TrLn \tag{49}
$$

Now $par(l, a)$ is a middle term in *TrLn* and the case is covered by **A**.

7. In case **D**, $par(l, a)$ is a term in a premiss of *SA*. We have two cases:

7.1. $par(l, a)$ is the removed term in *SA*:

$$
\cfrac{l' \parallel par(l, a) \quad \cfrac{c \in par(l, a) \quad \cfrac{c \in k \quad par(l, a) \parallel k}{par(l, a) = k}\ Unipar}{}\ SA}{l' \parallel k} \tag{50}
$$

We may assume that $par(l, a)$ is not in an equation in the derivation of the first premiss of *SA*. As in **6.3**, we get from that premiss and the third premiss of *Unipar* four cases: $l' = l$ and $l = k$ etc., and each leads to the elimination of *Unipar* concluding an equation with $par(l, a)$.

7.2. $par(l, a)$ is the right term in the second premiss of *SA*:

$$\frac{l' \parallel k \quad \dfrac{c \in k \quad c \in par(l, a) \quad k \parallel par(l, a)}{\dfrac{k = par(l, a)}{l' \parallel par(l, a)} \text{ } SA} \text{ } Unipar}{} \tag{51}$$

The third premiss of *Unipar* gives the two cases $k = l$ and $k \parallel l$, and these in combination with $l' \parallel k$ lead to $l' \parallel par(l, a)$ without rule *Unipar*.

The main part of the proof is now finished. In the proof for the projective case, it was sufficient to consider a line as a new term, the case of a point being dual. In affine geometry, new points can be introduced through rule *IA*, as in $a \in par(l, a)$. However, if a is a new term, also $par(l, a)$ is, and a cannot be a new term of maximal length. Our final task is therefore to ensure that the addition of the rules of affine geometry does not interfere with the proof of the subterm property for projective geometry. The essential case to consider is an instance of *Unipar* with a new term of maximal length $ln(a, b)$ in its conclusion, this being an uppermost occurrence of $ln(a, b)$ in an equation, as in

$$\frac{c \in ln(a, b) \quad c \in m \quad ln(a, b) \parallel m}{ln(a, b) = l} \text{ } Unipar \tag{52}$$

The third premiss has been concluded by *Sym*, *TrPar*, or *SA*, with $ln(a, b)$ always in a parallelism in a premiss. Only rule *Par* can introduce such a new term, but it cannot have an instance with $ln(a, b)$ a maximal term, so a term $ln(a, b)$ as in (52) is not a new term.   QED.

Lemma 10.6. *If the atomic cases Δ are derivable from the atomic assumptions Γ by rules I–V of plane affine geometry, they are already derivable by rules I–IV.*

Proof. As in the proof of Lemma 10.4, consider a first instance of rule *ET*. Rules *IA* and *Par* give possible new topformulas; however, if they contain any of x, y, z, or $ln(x, y)$, they also contain a *par*-construction with eigenvariables which would be a new term.   QED.

Lemmas 10.5 and 10.6 give us immediately the following theorem:

Theorem 10.6. *If the atomic cases Δ are derivable from the atomic assumptions Γ with rules I–V of plane projective geometry, there is a derivation in which all terms are terms in Γ, Δ.*

Derivability is decidable with a bounded algorithm, as in the case for projective geometry at the end of the previous section.

10.3 Examples of proof analysis in geometry

We give a couple of brief applications of proof analysis in geometry, to illustrate the control over the structure of derivations made possible by the restriction of proof search to known geometric objects.

1. Consistency. As a first application of the results of the previous sections, we obtain proofs of consistency. The standard formulation of consistency within sequent calculus is that the empty sequent $\rightarrow$ is underivable. Here we obtain the more general result that any finite set of atomic formulas is consistent:

Theorem 10.7. *If Γ contains only atoms, the sequent $\Gamma \rightarrow$ is not derivable in plane projective or affine geometry.*

Proof. If $\Gamma \rightarrow$ is derivable, it is derivable without rule *ET*, by Lemma 10.4 for projective and Lemma 10.7 for affine geometry. The remaining rules have always at least one formula as a conclusion.　　QED.

2. Euclid's fifth postulate. As a second application of the system of rules of affine geometry, we consider Euclid's fifth postulate: given a point a outside a line l, no point is incident with both l and the parallel to l through point a. Axiomatically, we may express this by the formula

$$\neg\, a \in l \supset\, \neg\, (b \in l\, \&\, b \in par(l, a))$$

It takes some effort to derive this formula from the axioms of affine geometry by standard methods of logical inference. Here, we can express the postulate as the 'logic-free' sequent

$$b \in l,\, b \in par(l, a) \rightarrow a \in l$$

Rule *Unipar* is essential in its derivation in our system:

Theorem 10.8. *If rule Unipar is deleted from the system of plane affine geometry and if the points a and b are not identical, the sequent*

$$b \in l,\, b \in par(l, a) \rightarrow a \in l$$

is not derivable.

Proof. No rule matches the premisses $b \in l,\, b \in par(l, a)$. The zero-premiss rule *Ref* produces loops, and *IA* and *Par* give $a \in par(l, a)$ and $par(l, a) \parallel l$. Now rule *Sym* gives $l \parallel par(l, a)$, and after it only loops are produced.　　QED.

With rule *Unipar* added, the proof search is straightforward: the two premisses $b \in par(l, a)$, $b \in l$ together with $par(l, a) \parallel l$ give $par(l, a) = l$, so $a \in par(l, a)$ gives $a \in l$ by line substitution.

Notes to Chapter 10

The first person to have attempted a combinatorial analysis of formal derivations in elementary geometry was Thoralf Skolem in 1920. The date is remarkable because systematic theories of the structure of proofs in mathematics were developed only from the 1930s on. Foremost among these are the sequent calculi and systems of natural deduction of Gentzen (1934–35).

Skolem's paper is famous for the Löwenheim–Skolem theorem, included in the first section. The second sections contains Skolem's result on lattice theory in a relational formulation, as presented in our Section 5.3.

Skolem used also in projective geometry a relational formulation of the theory and solved the derivability problem for an axiomatization that did not include non-collinearity. Instead of constructions, Skolem (1920) has axioms that guarantee the existence of connecting lines and of intersection points, such as the axiom $\forall x \forall y \exists z (x \in z \,\&\, y \in z)$. As noted in Section 8.2, a proof-theoretical analysis will not work if to such axioms is added an axiom of the form of non-collinearity. Skolem's main result was a solution to the word problem of the universal fragment of projective geometry. As an application, he gave a syntactic proof of the independence of the conjecture of Desargues.

Skolem's system of projective geometry was reformulated in terms of Gentzen's sequent calculus by Ketonen (1944). Ketonen also extended it to affine geometry, but again only as a universal theory, without the axiom of non-collinearity. A summary of Ketonen's thesis can be found in Bernays (1945), and a more detailed discussion in von Plato (2004). Ketonen's arguments are as hard to follow as are Skolem's. The geometrical parts of Skolem's 1920 paper and of Ketonen's thesis have remained completely unknown, even if both works otherwise have had a profound effect. Skolem's proof is analysed in von Plato (2007). Its starting point is the systematic development of proof systems for projective and affine geometry of von Plato (2010) (referred to as the manuscript 'Combinatorial Analysis of Proofs in Elementary Geometry' and with the year 2005 in the 2007 paper). It is shown in great detail how Skolem's result arises from the possibility of permuting the order of application of the geometrical axioms. The explicit notation for such permutations, as in this chapter, brought out more than one gap in the cases covered in Skolem's proof. Subsequently Bezem and Hendriks (2008) implemented a proof system for Skolem's geometry.

PART IV

Proof systems for non-classical logics

11 | Modal logic

11.1 The language and axioms of modal logic

In **modal logic**, we start from the language of propositional logic and add to it the two **modal operators** $\square$ and $\lozenge$, to form from any given formula A the formulas $\square A$ and $\lozenge A$. These are read as 'necessarily A' and 'possibly A', respectively.

A system of modal logic can be an extension of intuitionistic or classical propositional logic. In the latter, the notions of necessity and possibility are interdefinable by the equivalence $\square A \supset\subset \neg\lozenge\neg A$.

It is seen that necessity and possibility behave analogously to the quantifiers: in one interpretation, the necessity of A means that A holds in all circumstances, and the possibility of A means that A holds in some circumstances. The definability of possibility in terms of necessity is analogous to the classical definability of existence in terms of universality.

The system of **basic modal logic**, denoted by **K** in the literature, adds to the axioms of classical propositional logic the following:

Table 11.1 The system of basic modal logic

---

1. Axiom: $\square(A \supset B) \supset (\square A \supset \square B)$,
2. Rule of necessitation: from A to infer $\square A$.

One axiom and one rule is added to the axioms and rule *modus ponens* of propositional logic. The rule of necessitation requires that the premiss be derivable in the axiomatic system, i.e., its contents are that if A is a theorem, also $\square A$ is a theorem. The rule has caused considerable confusion in the literature.

If instead of axiomatic logic we start from a system of natural deduction for propositional logic, the following rules are added:

Table 11.2 Natural deduction for basic modal logic

$$\frac{\square(A \supset B) \quad \square A}{\square B} \qquad \frac{A}{\square A}$$

185

The second rule, called 'necessitation' or 'box introduction', requires a restriction. To see why, assume that from a given formula A one could conclude $\Box A$. Then, by first assuming A, then applying necessitation and implication introduction, one could conclude $A \supset \Box A$. Anything would imply its own necessity, which clearly is wrong. In the axiomatic formulation, the premiss of necessitation was a theorem because in axiomatic logic only theorems are derived. In a natural deduction system, one requires that A be derivable with no open assumptions. If one thinks of the analogy between necessity and universal quantification, it appears that the restriction is analogous to the variable condition in the rule for introducing the universal quantifier.

The analogy between necessity and possibility and the quantifiers suggests other operators similar to those of modal logic. For example, whatever must be done is **obligatory**, whatever can be done is **permitted**. These two notions belong to **deontic** logic. Even more simply, we can read $\Box A$ as 'always A' and $\Diamond A$ as 'some time A', respectively, which gives rise to **tense logic**.

The early study of modal logic, to the late 1950s, consisted mainly of suggested axiomatic systems based on an intuitive understanding of the basic notions. Certain axiomatizations became standard and are collected here in the form of a table. All of them start with the axioms of classical propositional logic and the axioms of basic modal logic of Table 11.1.

Table 11.3 Extensions of basic modal logic

	Axiom
T	$\Box A \supset A$
4	$\Box A \supset \Box \Box A$
E	$\Diamond A \supset \Box \Diamond A$
B	$A \supset \Box \Diamond A$
3	$\Box(\Box A \supset B) \vee \Box(\Box B \supset A)$
D	$\Box A \supset \Diamond A$
2	$\Diamond \Box A \supset \Box \Diamond A$
W	$\Box(\Box A \supset A) \supset \Box A$

Well-known extensions of basic modal logic are obtained through the addition of one or more of the above axioms to system **K**; for instance, **K4** is obtained by adding 4, **S4** by adding T and 4, **S5** by adding T, 4, and E (or T, 4, and B), deontic **S4** and **S5** are obtained by replacing axiom T with axiom D in **S4** and **S5**, respectively. The addition of W gives what is known as the Gödel–Löb system, **GL**. Axiom 2, also known as axiom M, gives the extensions of **K4** and **S4**, known as **K4.1** and **S4.1**, respectively. Axiom 3 is used for instance in the extension **S4.3** of system **S4**.

Of these systems, **GL** has been of particular interest for mathematical logic, because it encodes logical properties of the notion of **provability**. A very precise sense can be given to this encoding. Namely, it can be shown that the notion of derivability in a formal system of arithmetic, when internalized in arithmetic as in Gödel's incompleteness theorem, is captured by **GL**. An arithmetic **provability predicate** $\exists n Pr(n, m)$ expresses that there exists a Gödel number n of a formal derivation of the formula A with the Gödel number m. To the arithmetic notion of provability corresponds a modal operator $\Box A$ that expresses the provability of A. Any true arithmetic statement about provability is already derivable within the modal logic **GL**. This result about the completeness of provability logic is known as **Solovay's theorem**. Section 12.2 is devoted to provability logic.

The study of modal logic was completely changed in the late 1950s through the invention of a **relational semantics** of modal logic, to which we now turn.

11.2 Kripke semantics

What is known as Kripke semantics, also known as relational semantics, was presented by Saul Kripke in the late fifties (Kripke 1959) for the modal logic **S5**. It was modified later to accomodate also other modal logics (Kripke 1963a) and intuitionistic logic (Kripke 1965). The idea had several significant anticipations in the work of Arnould Bayart, Rudolf Carnap, Jaakko Hintikka, Stig Kanger, Richard Montague, Arthur Prior, and others. Questions of the originality and ultimate attribution for the invention of Kripke semantics have raised a considerable debate. We shall not take any position on these issues here, but refer to Copeland (2002) and Goldblatt (2005) for an in-depth discussion.

The basic idea of the semantics is that a proposition is necessary if and only if it is true in all 'possible worlds'. The idea is made precise as follows.

A **Kripke frame** is a set W, the elements of which are called **possible worlds**, together with an **accessibility relation** R, that is, a binary relation between elements of W. A Kripke frame becomes a **Kripke model** when a **valuation** is given. A valuation *val* takes a world w and an atomic formula

P and gives as value 0 or 1, to determine which atomic formulas are true at what particular worlds. The notation is

$$w \Vdash P \text{ whenever } val(w, P) = 1.$$

It is read as: *formula P is true at world w*, alternatively as *w forces P*. If $val(w, P) = 0$, we write $w \nVdash P$. A valuation is just like a line in a truth table, except that it is indexed by a world. If there is just one world, we have essentially the truth-table semantics of classical propositional logic. Valuations are supposed to be actually given, not just to exist in some abstract sense, so we have $w \Vdash P$ or $w \nVdash P$ for each atom P.

Valuations are extended in a unique way to arbitrary formulas by means of inductive clauses. For the propositional connectives, the inductive extension is straightforward:

Table 11.4 Valuations for the connectives

$w \Vdash A \& B$	whenever $w \Vdash A$ and $w \Vdash B$,
$w \Vdash A \lor B$	whenever $w \Vdash A$ or $w \Vdash B$,
$w \Vdash A \supset B$	whenever from $w \Vdash A$ follows $w \Vdash B$,
$w \Vdash \bot$	for no w.

It was assumed above that it is decidable if an atomic formula is forced at a given world. The same property holds then for arbitrary formulas, by the inductive clauses of Table 11.4. Further, if $w \nVdash A$, then $w \Vdash \neg A$. To prove this, assume $w \Vdash A$. A contradiction follows, so $w \Vdash \bot$. Therefore, by the inductive clause for implication, $w \Vdash \neg A$.

Definition 11.1. *Given a Kripke frame W, formula A is* **valid in** *W if, for every valuation, $w \Vdash A$ for every world w in W.*

The central idea in Kripke's semantics for modal logic is that a formula of the form $\Box A$ is true at world w if A is true at all worlds accessible from w through the relation R:

$$w \Vdash \Box A \text{ if and only if for all } o, \ o \Vdash A \text{ follows from } wRo.$$

The second key insight of Kripke semantics is that the axioms of different systems of modal logic correspond to special properties of the accessibility relation. Let us take what is probably the simplest example, namely a **reflexive** frame. We assume the accessibility relation to be reflexive. The condition corresponds to axiom T of Table 11.3:

$w \Vdash \Box A \supset A$ for every world w.

To see this, assume $w \Vdash \Box A$. Then $o \Vdash A$ for any o accessible from w, in particular, by reflexivity, for w itself, so $w \Vdash A$. Therefore $w \Vdash \Box A \supset A$. On the other hand, a frame that validates $\Box A \supset A$ has to be reflexive. To see this, let w be a world in the frame, and assume that not wRw. Then, under the valuation $v(A) \equiv \{o \in W | wRo\}$ we have the result that $w \Vdash \Box A$ but $w \nVdash A$. Thus reflexivity of the accessibility relation is equivalent to having a modal system with axiom T.

Similarly, it is seen that $\Box A \supset \Box \Box A$ is valid in every transitive frame and that every frame validating it has to be transitive. We say that there is a **correspondence** between a modal axiom and a property of the accessibility relation.

Observe that the defining axiom of the system of basic modal logic **K**, $\Box(A \supset B) \supset (\Box A \supset \Box B)$, is valid in every frame.

Table 11.6 of Section 11.3 gives a list of common modal axioms together with their corresponding frame conditions.

11.3 Formal Kripke semantics

Our aim is to provide a general approach to the proof theory of non-classical logics, through the use of **labelled** sequent calculi that are required to obey all the principles of good design usually required of traditional sequent systems. In particular, the calculi we shall present have all the structural rules – weakening, contraction, and cut – admissible; they support, whenever possible, proof search, and have a simple and uniform syntax that allows easy proofs of metatheoretical results.

In this section we shall present a sequent system for the basic modal logic **K** with rules for the modalities $\Box$ and $\Diamond$ obtained through a meaning explanation, in terms of the possible worlds semantics, and an inversion principle. The modal logic **K** is characterized by arbitrary frames. Restrictions of the class of frames that characterize a given modal logic amount to the addition of certain frame properties to the calculus. These properties are added in the form of mathematical rules, following the method of extension of sequent calculi presented in Chapter 6. All the extensions are thus obtained in a modular way. As a consequence, the structural properties of the resulting calculi can be established in one theorem for all systems.

(a) Basic modal logic. Basic modal logic is formulated as a labelled sequent calculus through an internalization of the possible worlds semantics within the syntax. The way to achieve the internalization is the following. First we enrich the language so that sequents are expressions of the form $\Gamma \rightarrow \Delta$ where the multisets Γ and Δ consist of **relational atoms** wRo and **labelled formulas** $w : A$, the latter corresponding to the forcing $w \Vdash A$ in Kripke models. Here w, o range over a set W of labels/possible worlds and A is any formula in the language of propositional logic extended by the modal operators of necessity and possibility, $\Box$ and $\Diamond$.

The rules for each connective and modality are obtained from their meaning explanations in terms of the relational semantics. Most importantly, the inductive definition of forcing for a modal formula is:

> $w \Vdash \Box A$ *whenever for all o, from wRo follows $o \Vdash A$.*

The definition gives:

> *If $o : A$ can be derived for an arbitrary o accessible from w, then*
>
> $w : \Box A$ *can be derived.*

This condition is turned into the formal rule

$$\frac{wRo, \Gamma \rightarrow \Delta, o : A}{\Gamma \rightarrow \Delta, w : \Box A} \; R\Box$$

In the rule, the arbitrariness of o becomes the variable condition that o must not occur in Γ, Δ.

Reading the semantical explanation in the other direction, we have the result that $w \Vdash \Box A$ and wRo give $o \Vdash A$. A corresponding rule for the antecedent side is:

$$\frac{o : A, w : \Box A, wRo, \Gamma \rightarrow \Delta}{w : \Box A, wRo, \Gamma \rightarrow \Delta} \; L\Box$$

The rules for $\Diamond$ are obtained similarly from the semantic explanation

> $w : \Diamond A$ *whenever for some o, wRo and $o : A$.*

The rules of sequent calculus for the propositional connectives are obtained from the inductive definition of forcing, as in Table 11.4. The result is a labelling of the active formulas with the same label in the premisses and conclusion of each rule of the calculus **G3c** of Table 6.6. The following sequent calculus **G3K** for basic modal logic is thus obtained:

Table 11.5 The sequent calculus **G3K**

Initial sequents

$$w : P, \Gamma \rightarrow \Delta, w : P \qquad\qquad wRo, \Gamma \rightarrow \Delta, wRo$$

Propositional rules

$$\frac{w : A, w : B, \Gamma \rightarrow \Delta}{w : A\&B, \Gamma \rightarrow \Delta} \; L\& \qquad\qquad \frac{\Gamma \rightarrow \Delta, w : A \quad \Gamma \rightarrow \Delta, w : B}{\Gamma \rightarrow \Delta, w : A\&B} \; R\&$$

$$\frac{w : A, \Gamma \rightarrow \Delta \quad w : B, \Gamma \rightarrow \Delta}{w : A \vee B, \Gamma \rightarrow \Delta} \; L\vee \qquad \frac{\Gamma \rightarrow \Delta, w : A, w : B}{\Gamma \rightarrow \Delta, w : A \vee B} \; R\vee$$

$$\frac{\Gamma \rightarrow \Delta, w : A \quad w : B, \Gamma \rightarrow \Delta}{w : A \supset B, \Gamma \rightarrow \Delta} \; L\supset \qquad \frac{w : A, \Gamma \rightarrow \Delta, w : B}{\Gamma \rightarrow \Delta, w : A \supset B} \; R\supset$$

$$\frac{}{w :\perp, \Gamma \rightarrow \Delta} \; L\perp$$

Modal rules

$$\frac{o : A, w : \Box A, wRo, \Gamma \rightarrow \Delta}{w : \Box A, wRo, \Gamma \rightarrow \Delta} \; L\Box \qquad \frac{wRo, \Gamma \rightarrow \Delta, o : A}{\Gamma \rightarrow \Delta, w : \Box A} \; R\Box$$

$$\frac{wRo, o : A, \Gamma \rightarrow \Delta}{w : \Diamond A, \Gamma \rightarrow \Delta} \; L\Diamond \qquad \frac{wRo, \Gamma \rightarrow \Delta, w : \Diamond A, o : A}{wRo, \Gamma \rightarrow \Delta, w : \Diamond A} \; R\Diamond$$

In the first initial sequent, P is an arbitrary atomic formula. In $R\Box$ and in $L\Diamond$, o is a fresh label. Observe that atoms of the form wRo in the right-hand side of sequents are never active in the logical rules nor in the rules that extend the logical calculus. Moreover, the derivations of the modal axioms that correspond to the properties of the accessibility relation do not use these sequents. As a consequence, initial sequents of the form $wRo, \Gamma \rightarrow \Delta, wRo$ are needed only for deriving properties of the accessibility relation, namely the axioms that correspond to the rules for R given below. Thus such initial sequents can as well be left out from the calculus without impairing the completeness of the system.

(b) Extensions. Our aim is to extend the above basic calculus so that the structural properties of the extensions are automatically guaranteed. These properties will follow from the form of the axioms that characterize the extensions. From the previous chapters, we know that they can be universal formulas, geometric formulas, or co-geometric formulas. The following table continues Table 11.3 with the frame properties of modal axioms:

Table 11.6 Modal axioms with corresponding frame properties

	Axiom	Frame property
T	$\Box A \supset A$	$\forall w\; wRw$ reflexivity
4	$\Box A \supset \Box\Box A$	$\forall wor(wRo \,\&\, oRr \supset wRr)$ transitivity
E	$\Diamond A \supset \Box\Diamond A$	$\forall wor(wRo \,\&\, wRr \supset oRr)$ euclideanness
B	$A \supset \Box\Diamond A$	$\forall wo(wRo \supset oRw)$ symmetry
3	$\Box(\Box A \supset B) \vee \Box(\Box B \supset A)$	$\forall wor(wRo \,\&\, wRr \supset oRr \vee rRo)$ connectedness
D	$\Box A \supset \Diamond A$	$\forall w\exists o\; wRo$ seriality
2	$\Diamond\Box A \supset \Box\Diamond A$	$\forall wor(wRo \,\&\, wRr \supset \exists l(oRl \,\&\, rRl))$ directedness
W	$\Box(\Box A \supset A) \supset \Box A$	no infinite R-chains $+$ transitivity

The frame properties in the first group (T, 4, E, B, 3) are universal axioms, those in the second group are geometric implications, as defined in Section 8.1, whereas the last one is not expressible as a first-order property.

The systems **T**, **K4**, **KB**, **S4**, **B**, **S5**, ... are obtained by adding one or more axioms to the system **K**. Sequent calculi are obtained by adding to the system **G3K** the rules that correspond to the properties of the accessibility relation that characterize their frames. For instance, a sequent calculus for **S4** is obtained by adding to **G3K** the rules that correspond to the axioms of reflexivity and transitivity of the accessibility relation:

$$\frac{wRw, \Gamma \to \Delta}{\Gamma \to \Delta}\ Ref \qquad\qquad \frac{wRr, wRo, oRr, \Gamma \to \Delta}{wRo, oRr, \Gamma \to \Delta}\ Trans$$

A system for **S5** is obtained by adding also the rule that corresponds to symmetry:

$$\frac{oRw, wRo, \Gamma \to \Delta}{wRo, \Gamma \to \Delta}\ Sym$$

Observe that rule *Trans*, as well as the rule that corresponds to euclideanness, have two principal atoms in the conclusion and are therefore subject to the closure condition.

$$\frac{oRr, wRo, wRr, \Gamma \to \Delta}{wRo, wRr, \Gamma \to \Delta}\ Eucl$$

The contracted instances of these rules are, respectively,

$$\frac{wRw, wRw, \Gamma \to \Delta}{wRw, \Gamma \to \Delta}\ Trans^{*} \qquad\qquad \frac{oRo, wRo, \Gamma \to \Delta}{wRo, \Gamma \to \Delta}\ Eucl^{*}$$

Both of the contracted rules are instances of rule *Ref*; therefore, in order to have the rule of contraction admissible, they have to be added into systems that do not contain rule *Ref.* Similar additions must be made for all extensions by rules that have instances with two occurrences of the same relational atom in the conclusion. A finer analysis shows that *Trans** is indeed admissible and need not be added.

Extensions are obtained in a modular way for all possible combinations of properties:

G3T = **G3K** + *Ref*
G3K4 = **G3K** + *Trans*
G3KB = **G3K** + *Sym*
G3S4 = **G3K** + *Ref* + *Trans*
G3TB = **G3K** + *Ref* + *Sym*
G3S5 = **G3K** + *Ref* + *Trans* + *Sym*

A system for **deontic logic** is obtained by the addition of the geometric rule *Ser*:

$$\frac{wRo, \Gamma \to \Delta}{\Gamma \to \Delta} \, Ser$$

Here the variable condition is $o \notin \Gamma, \Delta$.

Directedness is another property that follows the pattern of a geometric implication, and it is converted into the rule

$$\frac{oRl, rRl, wRo, wRr, \Gamma \to \Delta}{wRo, wRr, \Gamma \to \Delta} \, Dir$$

The variable condition is $l \notin wRo, wRr, \Gamma, \Delta$.

The treatment of a modal logic with a frame property not expressible as a first-order sentence, namely **provability logic**, is postponed to Section 12.2.

11.4 Structural properties of modal calculi

Let **G3K*** be any extension of **G3K** by rules for the accessibility relation that follow the rule scheme for extensions of sequent calculus (Table 6.10) or the more general geometric rule scheme (Table 8.1). The following properties can be established uniformly for all systems that belong to the class **G3K***.

Lemma 11.2. *Sequents of the form*

$$w : A, \Gamma \to \Delta, w : A$$

with A an arbitrary modal formula, are derivable in **G3K***.

Proof. By induction on the length of A.   QED.

To prove the correspondence between our systems and their Hilbert-style presentations, it is necessary to show that the characteristic axioms are derivable and the systems closed under the rules of necessitation and *modus ponens*.

Lemma 11.3. *For arbitrary A and B, the sequent*

$$\rightarrow w : \Box(A \supset B) \supset (\Box A \supset \Box B)$$

is derivable in **G3K***.

Proof. Apply, root first, the rules of **G3K** and Lemma 11.2.   QED.

The rule of necessitation with labels added is

$$\frac{\rightarrow w : A}{\rightarrow w : \Box A}$$

It is a context-dependent rule, as it requires both the antecedent and succedent contexts to be empty. As an explicit rule, it would impair the flexibility of the systems in the permutations that are needed for proving cut elimination; however, we do not need to add any such rule because we can show that it is admissible. To prove this, we exploit the first-order features of the system to show a lemma about substitution.

Substitution of labels is defined in the obvious way for relational atoms and labelled formulas:

$$wRo(r/l) \equiv wRo \text{ if } l \neq w \text{ and } l \neq o$$
$$wRo(r/w) \equiv rRo \text{ if } w \neq o$$
$$wRo(r/o) \equiv wRr \text{ if } w \neq o$$
$$wRw(r/w) \equiv rRr$$
$$w : A(r/o) \equiv w : A \text{ if } o \neq w$$
$$w : A(r/w) \equiv r : A$$

It is extended to multisets componentwise. We have

Lemma 11.4. *If $\Gamma \rightarrow \Delta$ is derivable in* **G3K***, *then $\Gamma(o/w) \rightarrow \Delta(o/w)$ is also derivable, with the same derivation height.*

Proof. By induction on the height n of the derivation of $\Gamma \rightarrow \Delta$.

If $n = 0$, and (o/w) is not a vacuous substitution, the sequent can either be an initial sequent of the form $w : P, \Gamma' \rightarrow \Delta', w : P$ or of the form $wRo, \Gamma' \rightarrow \Delta', wRo$ or a conclusion of $L\bot$ of the form $\bot, \Gamma' \rightarrow \Delta$. In each case $\Gamma(o/w) \rightarrow \Delta(o/w)$ is either an initial sequent of the same form or a conclusion of $L\perp$.

Suppose $n > 0$, and consider the last rule applied in the derivation. If it is a propositional rule, apply the inductive hypothesis to the premisses of the rule, and then the rule. Proceed similarly if the last rule is a modal rule without a variable condition, i.e., $L\square$ or $R\lozenge$. If the last rule is a modal rule with a variable condition, observe that either the substitution is vacuous or w is not an eigenvariable of the rule. In the first case, the result of the substitution is identical to $\Gamma \rightarrow \Delta$ and there is nothing to prove. In the second case, assume that o is not an eigenvariable. We have, in the case that the last rule is $R\square$ and $w : \square A$ appears as principal, a derivation that ends with

$$\frac{\vdots \\ wRr, \Gamma \rightarrow \Delta', r : A}{\Gamma \rightarrow \Delta', w : \square A} R\square$$

Here $r \neq w$ and r is not in Γ, Δ. By applying the inductive hypothesis to the shorter derivation of the premiss, and $R\square$, we obtain the derivation in n steps

$$\frac{\vdots \\ oRr, \Gamma(o/w) \rightarrow \Delta'(o/w), r : A}{\Gamma(o/w) \rightarrow \Delta'(o/w), o : \square A} R\square$$

If o is the eigenvariable, the derivation ends with

$$\frac{\vdots \\ wRo, \Gamma \rightarrow \Delta', o : A}{\Gamma \rightarrow \Delta', w : \square A} R\square$$

We apply first the inductive hypothesis to replace the eigenvariable o with a fresh label r. By the variable condition, the substitution does not affect Γ or Δ', and we obtain a derivation of height $n - 1$ of

$$wRr, \Gamma \rightarrow \Delta', r : A$$

Then we apply the inductive hypothesis to substitute w by o and conclude by $R\square$ in n steps

$$\frac{\vdots \\ oRr, \Gamma(o/w) \rightarrow \Delta'(o/w), r : A}{\Gamma(o/w) \rightarrow \Delta'(o/w), o : \square A} R\square$$

If w is not the label of a formula that is principal in the rule, the proof does not present any significant difference, and the case of $L\Diamond$ is detailed similarly.

For extensions of **G3K** with rules for the accessibility relation R, observe that the rules are schematic, thus closed under substitution. In other words, the induction proceeds as for the propositional rules.

For geometric extensions, some care is needed to avoid a clash with the eigenvariables of the geometric rule scheme. Suppose that the last rule in the derivation is one of the form

$$\frac{\overline{Q}_1(o_1/w_1), \overline{P}, \Gamma \rightarrow \Delta \quad \ldots \quad \overline{Q}_n(o_n/w_n), \overline{P}, \Gamma \rightarrow \Delta}{\overline{P}, \Gamma \rightarrow \Delta} \, GRS$$

If $o \neq o_i$ for all $i = 1, \ldots, n$, apply the inductive hypothesis to each of the premisses to obtain derivations of

$$\overline{Q}_i(o_i/w_i)(o/w), \overline{P}(o/w), \Gamma(o/w) \rightarrow \Delta(o/w)$$

Application of the geometric rule scheme gives

$$\overline{P}(o/w), \Gamma(o/w) \rightarrow \Delta(o/w)$$

If $o = o_i$ for some i, we replace first the eigenvariable o_i with a fresh variable o_i' by the inductive hypothesis applied to the i-th premiss of the rule. Then by the inductive hypothesis applied to each of the new premisses, we make the substitution o/w and obtain the conclusion by applying rule GRS.   QED.

Theorem 11.5. *The rules of weakening*

$$\frac{\Gamma \rightarrow \Delta}{w : A, \Gamma \rightarrow \Delta} \, LW \qquad \frac{\Gamma \rightarrow \Delta}{\Gamma \rightarrow \Delta, w : A} \, RW$$

$$\frac{\Gamma \rightarrow \Delta}{wRo, \Gamma \rightarrow \Delta} \, LW_R \qquad \frac{\Gamma \rightarrow \Delta}{\Gamma \rightarrow \Delta, wRo} \, RW_R$$

are height-preserving admissible in **G3K***.

Proof. Straightforward induction on the height of the derivation of the premiss for the propositional rules and the modal and non-logical rules without a variable condition. If the last step is a modal rule with a variable condition, the substitution lemma is applied to the premisses of the rule to have a fresh eigenvariable that does not clash with those in $w : A$ and wRo. The conclusion is then obtained by applying the inductive hypothesis and the modal rule. An identical procedure is applied if the last step is a geometric rule and $w : A$ or wRo contain some of its eigenvariables.   QED.

We are now ready to prove admissibility of the necessitation rule:

Corollary 11.6. *The necessitation rule is admissible in* **G3K***.

Proof. Suppose we have a derivation of $\rightarrow w : A$. By the substitution lemma, we obtain a derivation of $\rightarrow o : A$ and, by the admissibility of weakening, of $wRo \rightarrow o : A$. By $R\Box$, we have $\rightarrow w : \Box A$.   QED.

We also obtain a very useful property of a sequent calculus, namely:

Lemma 11.7. *All the rules of* **G3K*** *are height-preserving invertible.*

Proof. The proof of height-preserving invertibility for the propositional rules is done exactly as for **G3c** (theorem 3.1.1 in *Structural Proof Theory*). Rules $L\Box$ and $R\Diamond$ are trivially height-preserving invertible, because their premisses are obtained from the conclusion by weakening, and weakening is height-preserving invertible. The same holds for the rules for R. As usual, some care is needed for the rules with variable conditions.

We show the height-preserving invertibility of $R\Box$ by induction on the height n of the derivation of $\Gamma \rightarrow \Delta, w : \Box A$. If $n = 0$, it is an initial sequent or a conclusion of $L\perp$, but then also $wRo, \Gamma \rightarrow \Delta, o : A$ is an initial sequent or a conclusion of $L\perp$ (observe that it is essential here that the initial sequents are restricted to atomic formulas). If $n > 0$ and $\Gamma \rightarrow \Delta, w : \Box A$ is concluded by any rule $\mathcal{R}$ other than $R\Box$ or $L\Diamond$, we apply the inductive hypothesis to the premisses $\Gamma' \rightarrow \Delta', w : \Box A$ ($\Gamma'' \rightarrow \Delta'', w : \Box A$) and obtain derivations of height $n - 1$ of $wRo, \Gamma' \rightarrow \Delta', o : A$ ($wRo, \Gamma'' \rightarrow \Delta'', o : A$). By applying rule $\mathcal{R}$ we obtain a derivation of height n of $wRo, \Gamma \rightarrow \Delta, o : A$. If $\Gamma \rightarrow \Delta, w : \Box A$ is concluded by $L\Diamond$, Γ is of the form $r : \Diamond B, \Gamma'$ and the derivation ends with

$$\frac{rRl, l : B, \Gamma' \rightarrow \Delta, w : \Box A}{r : \Diamond B, \Gamma' \rightarrow \Delta, w : \Box A} \, L\Diamond$$

Here we can assume, without loss of generality, that the eigenvariable of $L\Diamond$ is not o (or else apply the substitution lemma). By the inductive hypothesis applied to the premiss, we obtain a derivation with the same derivation height that ends with

$$\frac{wRo, rRl, l : B, \Gamma' \rightarrow \Delta, o : A}{wRo, r : \Diamond B, \Gamma' \rightarrow \Delta, o : A} \, L\Diamond$$

If $\Gamma \rightarrow \Delta, w : \Box A$ is a conclusion of $R\Box$ with principal formula in Δ, we proceed in a similar way. If instead the principal formula is $\Box A$, the premiss of the last step gives the conclusion (possibly with a different eigenvariable, but the desired one can be obtained by height-preserving substitution). The proof of height-preserving invertibility of $L\Diamond$ is similar.   QED.

We are now in a position to prove the most important structural property of our calculi besides cut-admissibility, namely the height-preserving admissibility of contraction. First observe that there are, *a priori*, four contraction rules, left and right contraction for expressions of the form $w : A$ and of the form wRo. Explicitly stated, the rules of left and right contraction are:

$$\frac{w : A, w : A, \Gamma \to \Delta}{w : A, \Gamma \to \Delta} \, LC \qquad \frac{wRo, wRo, \Gamma \to \Delta}{wRo, \Gamma \to \Delta} \, LC_R$$

$$\frac{\Gamma \to \Delta, w : A, w : A}{\Gamma \to \Delta, w : A} \, RC \qquad \frac{\Gamma \to \Delta, wRo, wRo}{\Gamma \to \Delta, wRo} \, RC_R$$

Observe that rule RC_R is not needed if we use the calculus without the initial sequent $wRo, \Gamma \to \Delta, wRo$.

Theorem 11.8. *The rules of contraction are height-preserving admissible in* **G3K***.

Proof. By simultaneous induction on the height of derivation for left and right contraction.

If $n = 0$ the premiss is either an initial sequent or a conclusion of $L\perp$. In each case, the contracted sequent is also an initial sequent or a conclusion of $L\perp$.

If $n > 0$, consider the last rule used to derive the premiss of contraction. If the contraction formula is not principal in it, both occurrences are found in the premisses of the rule and they have a smaller derivation height. By the induction hypothesis, they can be contracted and the conclusion is obtained by applying the rule to the contracted premisses. If the contraction formula is principal in it, we distinguish three cases: 1. A rule in which the principal formulas appear also in the premiss (such as $L\Box$ or $R\Diamond$ or the rules for R). 2. A rule in which the active formulas are proper subformulas of the principal formula (such as the rules for $\&, \vee, \supset$; a formal definition of subformulas of a labelled formula is given in Section 11.5). 3. A rule in which active formulas are atoms wRo and proper subformulas of the principal formula (like rules $R\Box$ and $L\Diamond$).

In the first case we have, for instance,

$$\frac{w : \Box A, w : \Box A, wRo, o : A, \Gamma \to \Delta}{w : \Box A, w : \Box A, wRo, \Gamma \to \Delta} \, L\Box$$

By the induction hypothesis applied to the premiss we obtain

$$\frac{w : \Box A, wRo, o : A, \Gamma \to \Delta}{w : \Box A, wRo, \Gamma \to \Delta} \, L\Box$$

Observe that the case in which both contraction formulas are principal in a rule for R is taken care of by the closure condition.

In the second case, contraction is reduced to contraction on smaller formulas as in the standard proof for **G3c**.

In the third case, a subformula of the contraction formula and an atom wRo are found in the premiss, for instance

$$\frac{w : \Diamond A, wRo, o : A, \Gamma \to \Delta}{w : \Diamond A, w : \Diamond A, \Gamma \to \Delta} \, L\Diamond$$

By height-preserving invertibility applied to the premiss, we obtain a derivation of height $n - 1$ of

$$wRo, o : A, wRo, o : A, \Gamma \to \Delta$$

that yields, by the induction hypothesis for the two forms of contraction at left, a derivation of height $n - 1$ of

$$wRo, o : A, \Gamma \to \Delta$$

and the conclusion $w : \Diamond A, \Gamma \to \Delta$ follows in one more step by $L\Diamond$.

QED.

Also cut can take two forms, namely

$$\frac{\Gamma \to \Delta, w : A \quad w : A, \Gamma' \to \Delta'}{\Gamma, \Gamma' \to \Delta, \Delta'} \, Cut$$

and

$$\frac{\Gamma \to \Delta, wRo \quad wRo, \Gamma' \to \Delta'}{\Gamma, \Gamma' \to \Delta, \Delta'} \, Cut_R$$

However, Cut_R is not needed if the variant of **G3K** without the initial sequent $wRo, \Gamma \to \Delta, wRo$ is used.

We have:

Theorem 11.9. *The cut rule is admissible in* **G3K***.

Proof. The proof has the same structure as the proof of admissibility of cut for sequent calculus extended by the left rule scheme (Theorem 6.9). When the geometric rule scheme is considered, the proof follows the pattern of

Section 8.2, Theorem 8.10. We observe that in all the cases of permutation of cuts that may give a clash with the variable conditions in the modal rules (and in the rules for R in the case of geometric extensions), an appropriate substitution prior to the permutation, justified by Lemma 11.4, will be used.

We recall that the proof is by induction on the size of an uppermost cut formula in a derivation, with a subinduction on the height of cut, that is, the sum of the heights of the derivations of the premisses of cut. We consider in detail only the case of a cut with the cut formula principal in modal rules in both premisses of cuts.

If the cut formula is $w : \Box A$, the derivation is

$$\cfrac{\cfrac{wRo, \Gamma \rightarrow \Delta, o : A}{\Gamma \rightarrow \Delta, w : \Box A} \; R\Box \qquad \cfrac{wRr, w : \Box A, r : A, \Gamma' \rightarrow \Delta'}{wRr, w : \Box A, \Gamma' \rightarrow \Delta'} \; L\Box}{wRr, \Gamma, \Gamma' \rightarrow \Delta, \Delta'} \; Cut_1$$

It is transformed into

$$\cfrac{wRr, \Gamma \rightarrow \Delta, r : A \qquad \cfrac{\Gamma \rightarrow \Delta, w : \Box A \qquad wRr, w : \Box A, r : A, \Gamma' \rightarrow \Delta'}{wRr, r : A, \Gamma, \Gamma' \rightarrow \Delta, \Delta'} \; Cut_1}{\cfrac{wRr, wRr, \Gamma, \Gamma, \Gamma' \rightarrow \Delta, \Delta, \Delta'}{wRr, \Gamma, \Gamma' \rightarrow \Delta, \Delta'} \; Ctr^*} \; Cut_1$$

Here the upper cut is of smaller height and the lower is on a smaller cut formula, Ctr^* denotes repeated applications of contraction rules, and the leftmost premiss is obtained by the substitution (r/o) on the sequent $wRo, \Gamma \rightarrow \Delta, o : A$.

If the cut formula is $w : \Diamond A$, the derivation is

$$\cfrac{\cfrac{wRo, \Gamma \rightarrow \Delta, w : \Diamond A, o : A}{wRo, \Gamma \rightarrow \Delta, w : \Diamond A} \; R\Diamond \qquad \cfrac{wRr, r : A, \Gamma' \rightarrow \Delta'}{w : \Diamond A, \Gamma' \rightarrow \Delta'} \; L\Diamond}{wRo, \Gamma, \Gamma' \rightarrow \Delta, \Delta'} \; Cut_1$$

It is transformed into

$$\cfrac{\cfrac{wRo, \Gamma \rightarrow \Delta, w : \Diamond A, o : A \qquad w : \Diamond A, \Gamma' \rightarrow \Delta'}{wRo, \Gamma, \Gamma' \rightarrow \Delta, \Delta', o : A} \; Cut_1 \qquad wRo, o : A, \Gamma' \rightarrow \Delta'}{\cfrac{wRo, wRo, \Gamma, \Gamma', \Gamma' \rightarrow \Delta, \Delta', \Delta'}{wRo, \Gamma, \Gamma' \rightarrow \Delta, \Delta'} \; Ctr^*} \; Cut_1$$

Here the upper cut is of a smaller height and the lower on a smaller cut formula, and the rightmost premiss is obtained by the substitution (o/r) from $wRr, r : A, \Gamma' \rightarrow \Delta'$. QED.

11.5 Decidability

In general, cut elimination alone does not ensure a terminating proof search in a given system of sequent calculus. Cut elimination often has the subformula property as one of its immediate consequences. Sometimes the subformula property does not require full cut elimination, as in systems with 'analytic cut', i.e., with cut restricted to subformulas of the conclusion. Even the subformula property is not always sufficient to delimit the space of proof search, because the notion of a subformula can be extended, in first-order logic, to include all substitution instances of a given formula. Another source of possible non-termination is the presence of structural rules such as contraction.

In the systems we present, a suitable version of the subformula property, adequate for a proof of syntactic decidability, will emerge as a consequence of the structural properties of the calculi.

Before proceeding with the analysis of the subformula properties of our systems, we state precisely what we mean by a 'subformula' and a 'subformula property' of derivations in the context of prefixed formulas $w : A$.

Subformula. *For each propositional connective $\circ$, the subformulas of the formula $w : A \circ B$ are $w : A \circ B$ and all the subformulas of $w : A$ and of $w : B$. The subformulas of $w : \Box A$ and $w : \Diamond A$ are $w : \Box A$ and $w : \Diamond A$, respectively, and all the subformulas of $o : A$ for arbitrary o.*

Subformula property. *All formulas in a derivation are subformulas of formulas in the endsequent.*

Weak subformula property. *All formulas in a derivation are either subformulas of formulas in the endsequent or atomic formulas of the form wRo.*

A priori, these properties do not ensure decidability, unless a bound is found on the number of eigenvariables and of 'new worlds' in a derivation of a given sequent.

To obtain a bound on the number of atomic formulas that can appear in a derivation, it is useful to look at **minimal** derivations, that is, derivations in which shortenings are not possible. A derivation in which a rule instance produces, read in the root-first direction, a duplication of an atom wRo, can be shortened by the application of height-preserving admissibility of contraction at the rule instance. Similarly, a derivation that contains a sequent that matches the conclusion of a zero-premiss rule instance can be shortened by removing the subtree that had as its conclusion that sequent.

More precisely, we have:

Theorem 11.10. *All variables (worlds) in a minimal derivation of a sequent* $\Gamma \to \Delta$ *in* **G3K** *and in its extensions* **G3T, G3K4, G3KB, G3S4, G3TB,** *and* **G3S5** *are either eigenvariables or else variables in* Γ, Δ.

Proof. Immediate for **G3K** and its extensions with *Trans* and *Sym* (**G3K4,** **G3KB**). For extension with *Ref,* the proof follows from the lemma below.
                                                                                              QED.

Before stating the lemma, we observe that the hypothesis of minimality is redundant in the absence of *Ref.* Nevertheless, minimality is useful in any case because it precludes the possibility of applying rules that produce duplications.

Lemma 11.11. *All labels in atoms of the form* wRw *removed by Ref in a minimal derivation of a sequent* $\Gamma \to \Delta$ *in* **G3T, G3S4, G3TB, G3S5,** *are labels in* Γ, Δ.

Proof. Consider a minimal derivation of a sequent $\Gamma \to \Delta$ and suppose that there is a label w in an atom wRw removed by *Ref.* Consider a last occurrence of w and the step of *Ref* that removes it,

$$\frac{wRw, \Gamma \to \Delta}{\Gamma \to \Delta} \; Ref$$

Trace the atom wRw up in the derivation (observe that nothing, in particular no atom wRo, is removed by going up in the derivation).

If wRw is never principal in a rule, we trace it up to the leaves (initial sequents) of the derivation tree. If it is principal in an initial sequent, it has the form

$$wRw, \Gamma \to \Delta, wRw$$

We find an occurrence of w in the succedent. No atom of the form wRo is removed from the right-hand side of sequents in a derivation, so w is found in the conclusion. If wRw is not principal in any of the leaves, it can be removed altogether from the derivation, together with the instance of *Ref,* so the derivation is shortened, contrary to the assumption.

If wRw is principal in a rule, the rule is one of $L\Box$, $R\Diamond$, *Trans*, or *Sym*. We analyse each of these four possibilities.

1. If wRw is principal in $L\Box$, we have the derivation steps

$$\frac{\dfrac{w : A, wRw, w : \Box A, \Gamma' \to \Delta'}{wRw, w : \Box A, \Gamma' \to \Delta'} L\Box}{\begin{array}{c} \vdots\ \mathcal{D} \\ \dfrac{wRw, \Gamma \to \Delta}{\Gamma \to \Delta} Ref \end{array}}$$

By tracing the variable w, we find another occurrence of the variable in a modal expression $w : \Box A$. Since by hypothesis the premiss of *Ref* contains the last occurrence of w, the occurrence in $w : \Box A$ has been removed from the derivation before the step of *Ref*. The expression $w : \Box A$ can be active in propositional rules that either maintain w on the left-hand side of the sequent or that move it to the right-hand side. Eventually we find

$$wRw, w : B, \Gamma'' \to \Delta' \tag{1}$$

or

$$wRw, \Gamma'' \to \Delta'', w : B \tag{2}$$

Observe that, because of the variable condition, w cannot disappear from (1) by $L\Diamond$, nor from (2) by $R\Box$. If $w : B$ is active in $L\Box$ (1) or $R\Diamond$ (2), then we find another occurrence of w in an atom rRw in the conclusion of the rule. The atom rRw can be removed only by *Ref*, so we must have $r \equiv w$ and therefore, for each of the two alternatives

$$\frac{w : B, wRw, w : \Box B, \Gamma \to \Delta}{wRw, w : \Box B, \Gamma \to \Delta} L\Box \qquad \frac{wRw, \Gamma \to \Delta, w : \Diamond B, w : B}{wRw, \Gamma \to \Delta, w : \Diamond B} R\Diamond$$

Now we still have w in the conclusion.

2. If wRw is principal in $R\Diamond$, the analysis is similar.
3. If wRw is principal in *Trans*, we have the derivation

$$\frac{\dfrac{wRw, wRr, wRr, \Gamma' \to \Delta'}{wRw, wRr, \Gamma' \to \Delta'} Trans}{\begin{array}{c} \vdots \\ \dfrac{wRw, \Gamma \to \Delta}{\Gamma \to \Delta} Ref \end{array}}$$

By applying height-preserving admissibility of contraction to the premiss of *Trans*, we obtain a shorter derivation of the same endsequent, contrary to the assumption:

$$wRw, wRr, \Gamma' \rightarrow \Delta'$$
$$\vdots$$
$$\frac{wRw, \Gamma \rightarrow \Delta}{\Gamma \rightarrow \Delta} \, Ref$$

4. If wRw is principal in *Sym*, we have

$$\frac{wRw, wRr, \Gamma' \rightarrow \Delta'}{wRw, \Gamma' \rightarrow \Delta'} \, Sym$$
$$\vdots$$
$$\frac{wRw, \Gamma \rightarrow \Delta}{\Gamma \rightarrow \Delta} \, Ref$$

Again, by applying height-preserving admissibility of contraction as above, we obtain a shorter derivation of the same endsequent, with the step of *Sym* made superfluous, contrary to the assumption.   QED.

The property stated by the above result will be referred to in brief as the subterm property of a derivation:

Subterm property. *All terms (labels, worlds) in a derivation are either eigen-variables or terms (labels, worlds) in the conclusion.*

Proofs of the subterm property for systems for lattice theory and linear order have been obtained by similar methods in Chapters 5 and 7.

Another source of a potentially non-terminating proof search is the repetition of the principal formulas in the premisses of $L\square$ and $R\diamond$. By the following lemmas, it is enough to apply them only once on any given pair of principal formulas $wRo, w : \square A$ or $wRo, w : \diamond A$. First we prove that if there are two applications of $L\square$ or $R\diamond$ on the same pair of principal formulas and on the same branch of the derivation, such applications can be made consecutive by the permutation of rules:

Lemma 11.12. *Rule $L\square$ permutes down with respect to rules $L\&, R\&, L\vee,$ $R\vee, L\supset, R\supset, L\square,$ and $R\diamond$. It also permutes with respect to instances of $R\square,$ $L\diamond,$ and mathematical rules if the principal atom of $L\square$ is not active in them.*

Proof. The permutation is straightforward in the case of a one-premiss propositional rule. For instance, for $L\&$ we have:

$$\cfrac{\cfrac{o:A, w:\Box A, wRo, r:C, r:D, \Gamma \to \Delta}{w:\Box A, wRo, r:C, r:D, \Gamma \to \Delta}\, {\scriptstyle L\Box}}{w:\Box A, wRo, r:C\&D, \Gamma \to \Delta}\, {\scriptstyle L\&} \quad\rightsquigarrow\quad \cfrac{\cfrac{o:A, w:\Box A, wRo, r:C, r:D, \Gamma \to \Delta}{o:A, w:\Box A, wRo, r:C\&D, \Gamma \to \Delta}\, {\scriptstyle L\&}}{w:\Box A, wRo, r:C\&D, \Gamma \to \Delta}\, {\scriptstyle L\Box}$$

In the case of a two-premiss rule, use of height-preserving admissibility of weakening is needed; the derivation is, for instance,

$$\cfrac{\cfrac{o:A, w:\Box A, wRo, \Gamma \to \Delta, r:C}{w:\Box A, wRo, \Gamma \to \Delta, r:C}\, {\scriptstyle L\Box} \quad w:\Box A, wRo, \Gamma \to \Delta, r:D}{w:\Box A, wRo, \Gamma \to \Delta, r:C\&D}\, {\scriptstyle R\&}$$

It is transformed into

$$\cfrac{\cfrac{o:A, w:\Box A, wRo, \Gamma \to \Delta, r:C \quad \cfrac{w:\Box A, wRo, \Gamma \to \Delta, r:D}{o:A, w:\Box A, wRo, \Gamma \to \Delta, r:D}\, {\scriptstyle LW}}{o:A, w:\Box A, wRo, \Gamma \to \Delta, r:C\&D}\, {\scriptstyle R\&}}{w:\Box A, wRo, \Gamma \to \Delta, r:C\&D}\, {\scriptstyle L\Box}$$

Here the right premiss of $R\&$ is obtained by weakening from the right premiss of the given derivation.

The permutation is done similarly for the other propositional rules. The permutation is straightforward for the modal rules and the mathematical rules, because of the additional hypothesis of no clash of active or principal formulas. For instance, the permutation of $L\Box$ over $R\Box$ is as follows:

$$\cfrac{\cfrac{o:A, w:\Box A, wRo, rRl, \Gamma \to \Delta, l:B}{w:\Box A, wRo, rRl, \Gamma \to \Delta, l:B}\, {\scriptstyle L\Box}}{w:\Box A, wRo, \Gamma \to \Delta, r:\Box B}\, {\scriptstyle R\Box} \quad\rightsquigarrow\quad \cfrac{\cfrac{o:A, w:\Box A, wRo, rRl, \Gamma \to \Delta, l:B}{o:A, w:\Box A, wRo, \Gamma \to \Delta, r:\Box B}\, {\scriptstyle R\Box}}{w:\Box A, wRo, \Gamma \to \Delta, r:\Box B}\, {\scriptstyle L\Box}$$

$$\text{QED.}$$

A similar lemma holds, *mutatis mutandis*, for the dual case of rule $R\Diamond$:

Lemma 11.13. *Rule $R\Diamond$ permutes down with respect to rules $L\&$, $R\&$, $L\vee$, $R\vee$, $L\supset$, $R\supset$, $L\Box$, and $R\Diamond$. It also permutes with respect to instances of $R\Box$, $L\Diamond$, and mathematical rules if the principal atom of $R\Diamond$ is not active in them.*

Corollary 11.14. *In a minimal derivation in **G3K** and in any of its extensions with rules for R, rules $L\Box$ and $R\Diamond$ cannot be applied more than once on the same pair of principal formulas on any branch.*

Proof. Suppose we have, say, $L\Box$ applied twice on $w : \Box A, wRo$. Then the derivation contains the steps

$$\dfrac{o : A, w : \Box A, wRo, \Gamma' \rightarrow \Delta'}{w : \Box A, wRo, \Gamma' \rightarrow \Delta'} \, L\Box$$

$$\vdots$$

$$\dfrac{o : A, w : \Box A, wRo, \Gamma \rightarrow \Delta}{w : \Box A, wRo, \Gamma \rightarrow \Delta} \, L\Box$$

By permuting down $L\Box$ with respect to the steps in the dotted part of the derivation, we obtain a derivation of the same height that ends with

$$\dfrac{\dfrac{o : A, o : A, w : \Box A, wRo, \Gamma \rightarrow \Delta}{o : A, w : \Box A, wRo, \Gamma \rightarrow \Delta} \, L\Box}{w : \Box A, wRo, \Gamma \rightarrow \Delta} \, L\Box$$

By applying height-preserving contraction on $o : A$ in place of the upper $L\Box$, a shorter derivation is obtained, contrary to the assumption of minimality.   QED.

Decidability for the basic modal logic **K** is obtained in the strongest form of an effective bound on proof search in the system **G3K**:

Theorem 11.15. *The system* **G3K** *allows a terminating proof search.*

Proof. Consider any given sequent to be shown derivable. Apply, root first, any propositional rules and modal rules that match the conclusion. The propositional rules each reduce the complexity of the sequents. Rules $R\Box$ and $L\Diamond$ remove one modal operator and add an atomic relation; rules $L\Box$ and $R\Diamond$ increase the complexity. However, by the corollary above, rules $L\Box$ and $R\Diamond$, once applied on a given pair of formulas, need not be so applied again. Therefore the number of applications of $L\Box$ with principal formula $w : \Box A$ is bounded by the number of atoms of the form wRo that may appear on the left-hand side of sequents in the derivation. This number, in turn, is bounded by the number of existing atoms of that form and atoms that can be introduced by applications of $R\Box$ with the principal formula $w : \Box B$ or applications of $L\Diamond$ with principal formula $w : \Diamond B$. A similar bound holds for the number of applications of $R\Diamond$ on a given principal formula.   QED.

Explicit bounds are computed as follows. First define **negative** and **positive** parts of a sequent $\Gamma \rightarrow \Delta$ as the negative and positive parts of the formula $\&\Gamma \supset \vee\Delta$. For any given sequent, let $n(\Box)$ be the number of $\Box$ in the negative parts of the sequent; $p(\Box)$, the number of $\Box$ in the positive parts of the sequent; $n(\Diamond)$, the number of $\Diamond$ in the negative parts of the sequent; and $p(\Diamond)$, the number of $\Diamond$ in the positive parts of the sequent.

If the endsequent does not contain any atom wRo, the number of applications of $L\Box$ in a minimal derivation is bounded by

$$n(\Box)(p(\Box) + n(\Diamond))$$

If there are a atoms in the antecedent of the endsequent, it is bounded by

$$n(\Box)(p(\Box) + n(\Diamond) + a)$$

The number of applications of $R\Diamond$ is bounded, if there are no atoms wRo in the endsequent, by

$$p(\Diamond)(p(\Box) + n(\Diamond))$$

If there are a such atoms, the number of applications is bounded by

$$p(\Diamond)(p(\Box) + n(\Diamond) + a)$$

By a similar argument we have:

Theorem 11.16. *The system* **G3T** *allows a terminating proof search.*

Proof. First, observe that by the subterm property, reflexivity can be restricted to atoms wRw with w a world in the conclusion or an eigenvariable introduced by $R\Box$ or $L\Diamond$. Therefore, if e denotes the number of worlds in the endsequent, the bound to the number of applications of $L\Box$ and $R\Diamond$ is as above, with the parameter a replaced by $a + e + p(\Box) + n(\Diamond)$.   QED.

The addition of rule *Sym* to **G3K** or **G3T** has the following effect on proof search (with minimal derivations): whenever an atom wRo appears on the left-hand side of sequents, the symmetric atom oRw has to be added. If $w \equiv o$, no addition is needed, because such addition would cause a duplication and a use of height-preserving admissibility of contraction would shorten the derivation. With the notation introduced above, in **G3KB**, the bound to the number of applications of $L\Box$ is $n(\Box)(2p(\Box) + 2n(\Diamond) + 2a)$ and for $R\Diamond$, $p(\Diamond)(2p(\Box) + 2n(\Diamond) + 2a)$. For **G3TB**, the bounds are given by $n(\Box)(3p(\Box) + 3n(\Diamond) + 2a + e)$ and $p(\Diamond)(3p(\Box) + 3n(\Diamond) + 2a + e)$, respectively. We have thus proved:

Theorem 11.17. *The systems* **G3KB** *and* **G3TB** *allow a terminating proof search.*

In **G3S4**, the situation is more complicated. By the rule of transitivity and its interaction with $R\square$ that brings in new accessible worlds, we can build chains of accessible worlds on which $L\square$ can be applied *ad infinitum*. However, by our results on height-preserving admissibility of substitution and height-preserving admissibility of contraction, we can truncate an attempted proof search after a finite number of steps. Before giving precise bounds, we illustrate the method with an example (based upon a similar example discussed in section 11.2 of Viganó 2000). In what follows, we shall for simplicity restrict the language to the $\square$ modality. The results can be generalized by symmetry to the full language that includes $\diamond$.

We attempt to find a proof for the sequent $\rightarrow w : \square\neg\square A \supset \square B$. Proceeding root first, we build the following inference tree (in which we have omitted the derivable right premisses of $L\supset$):

$$\vdots$$

$$
\frac{\dfrac{\dfrac{\dfrac{\dfrac{\dfrac{\dfrac{\dfrac{rRl, wRr, wRo, oRr, w : \square\neg\square A \rightarrow o : B, r : A, l : A}{wRr, wRo, oRr, w : \square\neg\square A \rightarrow o : B, r : A, r : \square A}\,{\scriptstyle R\square}}{r : \neg\square A, wRr, wRo, oRr, w : \square\neg\square A \rightarrow o : B, r : A}\,{\scriptstyle L\supset}}{wRr, wRo, oRr, w : \square\neg\square A \rightarrow o : B, r : A}\,{\scriptstyle L\square}}{wRo, oRr, w : \square\neg\square A \rightarrow o : B, r : A}\,{\scriptstyle Trans}}{wRo, w : \square\neg\square A \rightarrow o : B, o : \square A}\,{\scriptstyle R\square}}{o : \neg\square A, wRo, w : \square\neg\square A \rightarrow o : B}\,{\scriptstyle L\supset}}{wRo, w : \square\neg\square A \rightarrow o : B}\,{\scriptstyle L\square}}{\dfrac{w : \square\neg\square A \rightarrow w : \square B}{\rightarrow w : \square\neg\square A \supset \square B}\,{\scriptstyle R\supset}}
$$

Consider now the topsequent. By applying the substitution (r/l) we obtain a derivation of the same height for the sequent

$$rRr, wRr, wRo, oRr, w : \square\neg\square A \rightarrow o : B, r : A, r : A$$

Hence, by height-preserving contraction, we have a derivation of

$$rRr, wRr, wRo, oRr, w : \square\neg\square A \rightarrow o : B, r : A$$

By a step of reflexivity we obtain a derivation of

$$wRw, wRo, oRr, w : \square\neg\square A \rightarrow o : B, r : A$$

There is a shortening by two steps of the original derivation. We can assume that the attempted proof search is for a minimal derivation, so we have a contradiction and the sequent is not derivable.

This argument can be formalized by giving a bound to the number of successive applications of $R\square$ with principal formula $w : \square A$ on successive worlds accessible from w. Intuitively, only those applications that contribute to the unfolding of all the boxed negative subformulas of the endsequent through steps of $L\square$ are needed. Additional steps are superfluous as they give rise to duplications (modulo substitution) as soon as the innermost boxed formula in a negative part has been reached, as shown in the above example.

Theorem 11.18. *In a minimal derivation of a sequent in* **G3S4**, *for each formula* $w : \square A$ *in its positive part there are at most* $n(\square)$ *applications of* $R\square$ *iterated on a chain of accessible worlds* wRw_1, $w_1Rw_2, \ldots$, *with principal formula* $w_i : \square A$.

Proof. Let m be $n(\square)$, and suppose that the antecedent of the derivable sequent contains a formula of the form $\square^m Q$ in which $\square^m$ denotes a block of m boxes. This assumption can be done without loss of generality: the modalities in the negative parts of the sequent do not necessarily occur in a block, but may be interleaved with propositional connectives. However, these connectives can be unfolded by the application, root first, of propositional rules without changing the number of applications of $R\square$ that are necessary to reach the innermost non-modal formula. Suppose that we iterate $R\square$ on a chain of accessible worlds $w_0Rw_1, \ldots$, etc., with $w_0 \equiv w$. After the first application of $R\square$, we have the accessibility w_0Rw_1 and application of $L\square$ produces an antecedent that contains $w_0 : \square^m Q$, $w_1 : \square^{m-1} Q$. After the second application we have the new accessibility w_1Rw_2, and, by transitivity, w_0Rw_2, and applications of $L\square$ add to the antecedent the formulas $w_2 : \square^{m-2} Q$, $w_2 : \square^{m-1} Q$. After m applications, the antecedent contains in addition $w_m : Q, \ldots, w_m : \square^{m-1} Q$ and the succedent $w_{m-1} : A$. If we apply $R\square$ one more time, by the newly available steps of $L\square$ licensed by the accessibility w_mRw_{m+1}, we add to the antecedent also the formulas $w_{m+1} : Q, \ldots, w_{m+1} : \square^{m-1} Q$. These latter steps are superfluous. By Lemma 11.3, we can make the height-preserving substitution (w_{m+1}/w_m) and by Theorem 11.8 eliminate all the duplications that arise, while maintaining the derivation height. By the single step of reflexivity that eliminates the atom w_mRw_m, we obtain a shorter derivation of the sequent reached after m steps of $R\square$.   QED.

We therefore have:

Corollary 11.19. *The system* **G3S4** *allows a terminating proof search.*

By the remarks before Theorem 11.17, the result above extends directly to **G3S5**.

11.6 Modal calculi with equality, undefinability results

The syntax for systems **G3K*** can be extended with equality. The treatment of equality as a left rule system, as in Section 6.3, is easily implemented in the context of labelled calculi. A contraction- and cut-free system **G3K-Eq** of modal logic with equality is obtained by adding to **G3K** the rules of reflexivity and Euclidean transitivity of equality, and rules of substitution of equals:

Table 11.7 Rules for the equality relation

$$\frac{w = w, \Gamma \to \Delta}{\Gamma \to \Delta} \, Eq\text{-}Ref \qquad \frac{o = r, w = o, w = r, \Gamma \to \Delta}{w = o, w = r, \Gamma \to \Delta} \, Eq\text{-}Trans$$

$$\frac{oRr, w = o, wRr, \Gamma \to \Delta}{w = o, wRr, \Gamma \to \Delta} \, Repl_{R_1} \qquad \frac{wRr, o = r, wRo, \Gamma \to \Delta}{o = r, wRo, \Gamma \to \Delta} \, Repl_{R_2}$$

$$\frac{o : A, w = o, w : A, \Gamma \to \Delta}{w = o, w : A, \Gamma \to \Delta} \, Repl$$

As in Section 6.3, it can be shown that rule *Repl* can be restricted to atomic formulas:

$$\frac{o : P, w = o, w : P, \Gamma \to \Delta}{w = o, w : P, \Gamma \to \Delta} \, Repl_{At}$$

Its general form for arbitrary formulas, rule *Repl* above, becomes admissible.

We observe by way of an example that the modal axiom

$$\Diamond(A \& \Box B) \supset \Box(A \lor \Diamond A \lor B)$$

that corresponds to the frame property

$$\forall wor(wRo \& wRr \supset r = o \lor rRo \lor oRr)$$

converts to the rule

$$\frac{r = o, wRo, wRr, \Gamma \to \Delta \quad rRo, wRo, wRr, \Gamma \to \Delta \quad oRr, wRo, wRr, \Gamma \to \Delta}{wRo, wRr, \Gamma \to \Delta}$$

The corresponding sequent system is obtained by adding the above rule to the system **G3K** augmented with the rules for equality. All the structural properties of the resulting system hold, as a consequence of the general results.

The use of proof systems that unify the syntax and semantics of modal logic permits us to obtain very simple proofs of negative results in **modal correspondence theory**. These results state that certain frame properties, such as irreflexivity and intransitivity, do not have any modal correspondent. The usual proofs are based on model extension methods: to prove that a frame property is not modally definable, it is shown that the corresponding class of frames is not closed under the constructions of disjoint union, generated subframes, bounded morphic images, and ultrafilter extensions (cf. Blackburn, de Rijke, and Venema 2001, section 3.3; see also Van Benthem 1984). In our systems, the lack of a modal correspondent is an immediate consequence of a conservativity theorem. Consider, for example, the frame property of irreflexivity $\forall w \neg\, wRw$. It corresponds to the rule

$$\frac{}{wRw, \Gamma \to \Delta}\, Irref$$

We have by a straightforward proof analysis:

Theorem 11.20. *The system* **G3K**+Irref *is conservative over* **G3K**.

Proof. Suppose that the sequent $\Gamma \to \Delta$, not containing relational atoms, is derivable in **G3K**+*Irref*. The atoms of the form wRo that appear on the left-hand side of sequents in the derivation originate from applications of rule $R\Box$. By the variable condition, $w \neq o$, so the derivation contains no atom of the form wRw, hence no application of *Irref*. Therefore the sequent is derivable in **G3K**.   QED.

It follows that the property of irreflexivity does not have any modal correspondent, because if it had, there would be some formula that is provable in the extension **G3K**+*Irref* but not in **G3K**.

Intransitivity is given by the axiom

$$\forall w \forall o \forall r (wRo \& oRr \supset \neg wRr).$$

It corresponds to the zero-premiss rule

$$\frac{}{wRo, oRr, wRr, \Gamma \to \Delta}\, Intrans$$

A similar result obtains:

Theorem 11.21. *The system* **G3K**+Intrans *is conservative over* **G3K**.

Proof. As above, observe that relational atoms on the left in derivations of a sequent $\Gamma \to \Delta$ originate from applications of $R\Box$. In order to have both wRr and oRr, two applications of $R\Box$ with the same eigenvariable are needed, but this is ruled out by the variable condition.   QED.

We can generalize the above two results as follows:

Theorem 11.22. *Let* $P_1, \ldots, P_n\Gamma \to \Delta$ *be a zero-premiss rule, called G-Intrans, that corresponds to the axiom* $\neg(P_1 \& \ldots \& P_n)$ *with* $P_i \equiv w_i R o_i$, *and assume that for some* $i, j, o_i = o_j$. *Then* **G3K**+G-Intrans *is conservative over* **G3K**.

Proof. Straightforward.   QED.

By similar arguments, we can prove conservativity for extensions by rules for geometric axioms, such as the property that there exists a reflexive world, $\exists w\, wRw$, or compositions thereof, such as the property by which every world has access to a reflexive one, $\forall w\exists o(wRo \ \& \ oRo)$.

Let *Eref* be the geometric rule by which there exists a reflexive world

$$\frac{wRw, \Gamma \to \Delta}{\Gamma \to \Delta}$$

The variable condition is that w is not in Γ, Δ. We have

Theorem 11.23. *The system* **G3K**+Eref *is conservative over* **G3K**.

Proof. Assume that $\Gamma \to \Delta$ is derivable in **G3K**+*Eref* and consider a step of *Eref* in the derivation. Trace its active atom wRw until it is principal in a rule. The rule can be $L\Box$ or $R\Diamond$. In the former case, the derivation above the step of *Eref* contains a sequent of the form $wRw, w : \Box A, \Gamma' \to \Delta'$. By the variable condition on *Eref*, the label w in $w : \Box A$ has to disappear before the application of *Eref*. However, by the presence of w in wRw in the context, such a step would not be correct. The other possibility, that wRw is principal in $R\Diamond$, is excluded in a similar way. If wRw is principal in an initial sequent, then wRw is found in the succedent Δ because no relational atoms disappear from the right-hand side of sequents, but this violates the variable condition on *Eref*. The only possibility that remains is that the atom wRw is nowhere principal. Then we can remove it everywhere from the derivation, together with the step of *Eref*.

This procedure, combined with an induction on the number of occurrences of *Eref* in the derivation, produces a derivation in **G3K**.   QED.

Corollary 11.24. *The frame properties of irreflexivity, intransitivity and its generalization, and the existence of a reflexive world do not have any modal correspondent.*

Proof. By the conservativity theorems, there is no modal formula that can be proved in the systems extended by the above frame properties that could not be proved in the ground system.   QED.

11.7 Completeness

Kripke's original proof of completeness for modal logic with respect to the semantics bearing his name used a direct construction of a Beth tree from a failed proof search. In later proofs, Kripke countermodels had nodes built from Henkin sets of formulas and extra devices that impose additional properties on the accessibility relation. We show that for the labelled calculus introduced in the previous sections, a completeness proof can be given that is closer to Kripke's original argument. For every sequent, the proof search either ends in a proof or fails, and the failed proof tree gives a Kripke countermodel.

(a) Soundness. We first reformulate the semantical notions of Section 11.2 so that they apply to our labelled calculi:

Definition 11.25. *Let K be a frame with an accessibility relation $\mathcal{R}$ that satisfies the properties $*$. Let W be the set of variables (labels) used in derivations in* **G3K***. *An* **interpretation** *of the labels W in frame K is a function $[[\cdot]]$: $W \rightarrow K$. A* **valuation** *of atomic formulas in frame K is a map $\mathcal{V} : AtFrm \rightarrow \mathcal{P}(K)$ that assigns to each atom P the set of nodes of K in which P holds; the standard notation for $k \in \mathcal{V}(P)$ is $k \Vdash P$.*

Valuations are extended to arbitrary formulas by the following inductive clauses:

$k \Vdash \bot$ for no k,
$k \Vdash A \& B$ if $k \Vdash A$ and $k \Vdash B$,
$k \Vdash A \vee B$ if $k \Vdash A$ or $k \Vdash B$,
$k \Vdash A \supset B$ if from $k \Vdash A$ follows $k \Vdash B$,
$k \Vdash \Box A$ if for all k', from $k\mathcal{R}k'$ follows $k' \Vdash A$,
$k \Vdash \Diamond A$ if there exists k' such that $k\mathcal{R}k'$ and $k' \Vdash A$.

Definition 11.26. *A sequent $\Gamma \rightarrow \Delta$ is* **valid for an interpretation and a valuation** *in K if for all labelled formulas $w : A$ and relational atoms oRr in Γ, whenever $[[w]] \Vdash A$ and $[[o]]\mathcal{R}[[r]]$ in K, then for some $l : B$ in Δ, $[[l]] \Vdash B$. A sequent is* **valid** *if it is valid for every interpretation and every valuation in a frame.*

Theorem 11.27. *If the sequent* $\Gamma \rightarrow \Delta$ *is derivable in* **G3K***, then it is valid in every frame with the properties* *.

Proof. By induction on the derivation of $\Gamma \rightarrow \Delta$ in **G3K***. If it is an initial sequent, then there is a labelled atom $w : P$ both in Γ and in Δ, so the claim is obvious, and similarly if the sequent is a conclusion of $L\perp$, because for no valuation can $\perp$ be forced at any node.

If $\Gamma \rightarrow \Delta$ is a conclusion of a propositional rule, assume the rule is $L\&$ with the premiss $w : A, w : B, \Gamma' \rightarrow \Delta$. Assume that for an arbitrary assignment and interpretation, all the formulas in Γ are valid. Since the condition $[[w]] \Vdash A\&B$ is equivalent to $[[w]] \Vdash A$ and $[[w]] \Vdash B$, the inductive hypothesis, i.e., validity of $w : A, w : B, \Gamma' \rightarrow \Delta$ for every interpretation, gives the desired conclusion.

If $\Gamma \rightarrow \Delta$ is a conclusion of a modal rule, say $R\square$, with the premiss $wRo, \Gamma' \rightarrow \Delta', o : A$, assume by the induction hypothesis that the premiss is valid. Let $[[\cdot]]$ be an arbitrary interpretation that validates all the formulas in Γ'. We claim that one of the formulas in Δ' or $w : \square A$ is valid under this interpretation. Let k be an arbitrary element of K such that $[[w]]Rk$ holds in K; let $[[\cdot]]'$ be the interpretation identical to $[[\cdot]]$ except possibly on o, where we set $[[o]]' \equiv k$. Clearly $[[\cdot]]'$ validates all the formulas in the antecedent of the premiss, so it validates a formula in Δ' or $o : A$ (the alternative being independent of the choice of $[[o]]'$). In the former case we have the result that also $[[\cdot]]$ validates a formula in Δ'; in the latter, that $[[\cdot]]$ validates $w : \square A$.

If the sequent is a conclusion of a mathematical rule without eigenvariables, let the rule be for instance *Trans*:

$$\frac{wRr, wRo, oRr, \Gamma \rightarrow \Delta}{wRo, oRr, \Gamma \rightarrow \Delta}$$

Let $[[w]]\mathcal{R}[[o]]$ and $[[o]]\mathcal{R}[[z]]$. Since $\mathcal{R}$ satisfies transitivity by assumption, we have $[[w]]\mathcal{R}[[z]]$, so validity of the premiss gives the conclusion.

If the sequent is a conclusion of a mathematical rule with eigenvariables, let the rule be for instance *Directedness*:

$$\frac{oRl, rRl, wRo, wRr, \Gamma \rightarrow \Delta}{wRo, wRr, \Gamma \rightarrow \Delta}$$

Here l is an eigenvariable. Since by hypothesis the frame is directed, if $[[w]]\mathcal{R}[[o]]$ and $[[w]]\mathcal{R}[[r]]$, there exists d such that $[[o]]\mathcal{R}d$ and $[[r]]\mathcal{R}d$. The premiss is valid for all interpretations, in particular for one that coincides with $[[\cdot]]$ on all labels, except possibly on l where it is assigned value d (this choice is possible because l is an eigenvariable). It follows that one of the formulas in Δ holds under this interpretation.   QED.

(b) Completeness. The proof of completeness follows the pattern of proof of completeness for predicate logic, as in *Structural Proof Theory*, section 4.4.

Theorem 11.28. *Let $\Gamma \rightarrow \Delta$ be a sequent in the language of* **G3K***. *Then either the sequent is derivable in* **G3K*** *or it has a Kripke countermodel with properties* ∗.

Proof. We define for an arbitrary sequent $\Gamma \rightarrow \Delta$ in the language of **G3K*** a reduction tree, by applying the rules of **G3K***, root first, in all possible ways. If the construction terminates, we obtain a proof, or else the tree becomes infinite. By König's lemma an infinite tree has an infinite branch that is used to define a countermodel to the endsequent.

1. *Construction of the reduction tree.* The reduction tree is defined inductively in stages as follows:

Stage 0 has $\Gamma \rightarrow \Delta$ at the root of the tree. Stage $n > 0$ has two cases:

Case I: If every topmost sequent is an initial sequent or a conclusion of $L\bot$ or of a zero-premiss mathematical rule, the construction of the tree ends.

Case II: If this is not the case, we continue the construction of the tree by writing above those topsequents that are not initial, nor conclusions of $L\bot$ or of a zero-premiss mathematical rule, other sequents that are obtained by applying, root first, the rules of **G3K*** whenever possible, in a given order.

There are $10 + r$ different stages, 10 for the rules of the basic modal system, r for the mathematical rules. At stage $n = 10 + r + 1$, we repeat stage 1; at stage $n = 10 + r + 1$, we repeat stage 2; and so on, for each n.

We start, for $n = 1$, with $L\&$: consider topmost sequents of the form

$$w_1 : B_1 \& C_1, \ldots, w_m : B_m \& C_m, \Gamma' \rightarrow \Delta$$

Here $B_1 \& C_1, \ldots, B_m \& C_m$ are all the formulas in Γ with a conjunction as the outermost logical connective; we write

$$w_1 : B_1, w_1 : C_1, \ldots, w_m : B_m, w_m : C_m, \Gamma' \rightarrow \Delta$$

on top of it. This step corresponds to applying, root first, m times, rule $L\&$.

For $n = 2$, we consider all the sequents of the form

$$\Gamma \rightarrow w_1 : B_1 \& C_1, \ldots, w_m : B_m \& C_m, \Delta'$$

Here $w_1 : B_1 \& C_1, \ldots, w_m : B_m \& C_m$ are all the labelled formulas in the succedent with a conjunction as the outermost logical connective. We write on top of them the 2^m sequents

$$\Gamma \to w_1 : D_1, \ldots, w_m : D_m, \Delta'$$

Here D_i is either B_i or C_i and all possible choices are made. This step is equivalent to applying $R\&$, root first, successively with principal labelled formulas $w_1 : B_1 \& C_1, \ldots, w_m : B_m \& C_m$.

For $n = 3$ and 4 we consider $L\vee$ and $R\vee$ and define the reductions symmetrically to the cases $n = 2$ and $n = 1$, respectively.

For $n = 5$, for each topmost sequent that has a number of labelled formulas $w_1 : B_1 \supset C_1, \ldots, w_m : B_m \supset C_m$ with implication as the outermost logical connective in the antecedent, Γ' the other formulas, and succedent Δ, write on top of it the 2^m sequents

$$w_{i_1} : C_{i_1}, \ldots, w_{i_k} : C_{i_k}, \Gamma' \to w_{j_{k+1}} : B_{j_{k+1}}, \ldots, w_{j_m} : B_{j_m}, \Delta$$

Here $i_1, \ldots, i_k \in \{1, \ldots, m\}$ and $j_{k+1}, \ldots, j_m \in \{1, \ldots, m\} - \{i_1, \ldots, i_k\}$. This step, perhaps less transparent because of the double indexing, corresponds to the root-first application of rule $L\supset$ with principal formulas $w_1 : B_1 \supset C_1, \ldots, w_m : B_m \supset C_m$.

For $n = 6$, we consider all the labelled sequents that have implications in the succedent, say $w_1 : B_1 \supset C_1, \ldots, w_m : B_m \supset C_m$, and Δ' the other formulas, and write on top of them

$$w_1 : B_1, \ldots, w_m : B_m, \Gamma \to w_1 : C_1, \ldots, w_m : C_m, \Delta'$$

So here we apply, root first, m times, rule $R\supset$.

For $n = 7$, we consider all topsequents with modal formulas as in $w_1 : \Box B_1, \ldots, w_m : \Box B_m$ and relational atoms $w_1 R o_1, \ldots, w_m R o_m$ in the antecedent, and write on top of these sequents the sequents

$$o_1 : B_1, \ldots, o_m : B_m, w_1 : \Box B_1, \ldots,$$

$$w_m : \Box B_m, w_1 R o_1, \ldots, w_m R o_m, \Gamma' \to \Delta$$

Here we apply, m times, rule $L\Box$.

For $n = 8$, let $w_1 : \Box B_1, \ldots, w_m : \Box B_m$ be all the formulas with $\Box$ as the outermost connective in the succedent of topsequents of the tree, and let Δ'

be the other formulas. Let $r_1, \ldots, r_m$ be fresh variables, not yet used in the reduction tree, and write on top of each sequent the sequent

$$w_1 R r_1, \ldots, w_m R r_m, \Gamma \rightarrow \Delta', r_1 : B_1, \ldots, r_m : B_m$$

Here we apply, m times, rule $R\square$.

For $n = 9$, let $w_1 : \Diamond B_1, \ldots, w_m : \Diamond B_m$ be all the formulas with $\Diamond$ as the outermost connective in the antecedent of topsequents of the tree, and let Γ' be the other formulas. Let $l_1, \ldots, l_m$ be fresh variables, and write on top of each sequent the sequent

$$w_1 R l_1, \ldots, w_m R l_m, l_1 : B_1, \ldots, l_m : B_m \Gamma' \rightarrow \Delta$$

We apply, m times, rule $L\Diamond$.

For $n = 10$, consider all topsequents with modal formulas $w_1 : \Diamond B_1, \ldots,$ $w_m : \Diamond B_m$ in the succedent and relational atoms $w_1 R o_1, \ldots, w_m R o_m$ in the antecedent, and write on top of these sequents the sequents

$$w_1 R o_1, \ldots, w_m R o_m, \Gamma \rightarrow \Delta', w_1 : \Diamond B_1, \ldots,$$

$$w_m : \Diamond B_m, o_1 : B_1, \ldots, o_m : B_m$$

We apply, m times, rule $R\Diamond$.

Finally, for $n = 10 + j$, we consider the generic case of a mathematical rule, that is, a rule for the relation R. For systems with the subterm property, the mathematical rules need to be instantiated only on terms in the conclusion or on eigenvariables. Thus, if the system contains rule *Ref*, instances of that rule consist in adding to the antecedent all the relational atoms wRw for w in $\Gamma \rightarrow \Delta$; with a rule with eigenvariables, such as seriality, the step for that rule adds all the atoms of the form wRo for w in $\Gamma \rightarrow \Delta$ and o a fresh variable. Observe that because of height-preserving substitution and height-preserving admissibility of contraction, once a rule with eigenvariables has been considered, it need not be instantiated again on the same principal formulas. If it is a rule such as *Trans*, consider all the sequents with a pair of atoms of the form wRo, oRr in the antecedent and write on top of them the sequents with the atoms wRr added.

For any n, for sequents that are neither initial, nor conclusions of $L\bot$, nor of zero-premiss mathematical rules, nor treatable by any one of the above reductions, we write the sequent itself above them. This step is needed to treat uniformly the failure of proof search in the following two cases: the case in which the search goes on for ever because new rules always become applicable and the case in which a sequent is reached which is not a conclusion of any rule nor an initial sequent.

If the reduction tree is finite, all its leaves are initial or conclusions of $L \perp$, or of zero-premiss mathematical rules, and the tree, read from the leaves to the root, yields a derivation.

2. *Construction of the countermodel.* If the reduction tree is infinite, it has an infinite branch. Let $\Gamma_0 \to \Delta_0 \equiv \Gamma \to \Delta, \Gamma_1 \to \Delta_1 \ldots, \Gamma_i \to \Delta_i, \ldots$ be one such branch. Consider the sets of labelled formulas and relational atoms

$$\Gamma \equiv \bigcup_{i>0} \Gamma_i \qquad \Delta \equiv \bigcup_{i>0} \Delta_i$$

We define a Kripke model that forces all the formulas in Γ and no formula in Δ and is therefore a countermodel to the sequent $\Gamma \to \Delta$.

Consider the frame K the nodes of which are all the labels that appear in the relational atoms in Γ, with their mutual relationships expressed by the relations wRo in Γ. Clearly, the construction of the reduction tree imposes the frame properties of the countermodel; for instance, in the system **G3S4**, the constructed frame is reflexive and transitive. The model is defined as follows: for all atomic formulas $w : P$ in Γ, we stipulate that $w \Vdash P$ in the frame, and for all atomic formulas $o : Q$ in Δ we stipulate that $o \nVdash Q$. Since no sequent in the reduction tree is initial, this choice can be coherently made, for if there were the same labelled atom in Γ and in Δ, then, since the sequents in the reduction tree are defined in a cumulative way, for some i there would be a labelled atom $w : P$ both in the antecedent and in the succedent of $\Gamma_i \to \Delta_i$.

We then show inductively on the weight of formulas that A is forced in the model at node w if $w : A$ is in Γ and A is not forced at node w if $w : A$ is in Δ. Therefore we have a countermodel to the endsequent $\Gamma \to \Delta$.

If A is $\perp$, it cannot be in Γ because no sequent in the branch contains $w : \perp$ in the antecedent, so it is not forced at any node of the model.

If A is atomic, the claim holds by the definition of the model.

If $w : A \equiv w : B \& C$ is in Γ, there exists i such that $w : A$ appears first in Γ_i, and therefore, for some $l \geqslant 0$, $w : B$ and $w : C$ are in Γ_{i+l}. By the induction hypothesis, $w \Vdash B$ and $w \Vdash C$, and therefore $w \Vdash B \& C$.

If $w : A \equiv w : B \& C$ is in Δ, consider the step i in which the reduction for A applies. This gives a branching, and one of the two branches belongs to the infinite branch, so either $w : B$ or $w : C$ is in Δ, and therefore by the inductive hypothesis, $w \nVdash B$ or $w \nVdash C$, and therefore $w \nVdash B \& C$.

If $w : A \equiv w : B \lor C$ is in Γ, we reason similarly to the case of $w : A \equiv w : B \& C$ in Δ.

If $w : A \equiv w : B \lor C$ is in Δ, we argue as with $w : A \equiv w : B \lor C$ in Γ.

If $w : A \equiv w : B \supset C$ is in Γ, then either $w : B$ is in Δ or $w : C$ is in Γ. By the inductive hypothesis, in the former case $w \Vdash\!\!\!/ \; B$, and in the latter $w \Vdash C$, so in both cases $w \Vdash B \supset C$.

If $w : A \equiv w : B \supset C$ is in Δ, then for some i, $w : B \in \Gamma_i$ and $w : C \in \Delta_i$, so by the inductive hypothesis $w \Vdash B$ and $w \Vdash\!\!\!/ \; C$, therefore $w \Vdash\!\!\!/ \; B \supset C$.

If $w : A \equiv w : \Box B$ is in Γ, we consider all the relational atoms wRo that occur in Γ. If there is no such atom, then the condition that for all o accessible from w in the frame $o \Vdash B$, is vacuously satisfied, and therefore $w \Vdash \Box B$ in the model. Otherwise for any occurrence of wRo in Γ we find, by the construction of the reduction tree, an occurrence of $o : B$ in Γ. By the inductive hypothesis, $o \Vdash B$, and therefore $w \Vdash \Box B$ in the model.

If $w : A \equiv w : \Box B$ is in Δ, consider the step at which the reduction for $w : A$ applies. We then find $o : B$ in Δ for some o with wRo in Γ. By the induction hypothesis, $o \Vdash\!\!\!/ \; B$, and therefore $w \Vdash\!\!\!/ \; A$.

The cases of $w : A \equiv w : \Diamond B$ in Γ and of $w : A \equiv w : \Diamond B$ in Δ are symmetric to those of $w : A \equiv w : \Box B$ in Δ and of $w : A \equiv w : \Box B$ in Γ, respectively.   QED.

Corollary 11.29. *If a sequent $\Gamma \rightarrow \Delta$ is valid in every Kripke model with the frame properties $*$, then it is derivable in the system* **G3K*** .

Notes to Chapter 11

Various sources in the literature claim that the deduction theorem does not hold for modal logic. The reason for the claims of failure lies in an unrestricted necessitation rule, used for extending derivability in a Hilbert system to a notion of derivability from assumptions. The rule licences the derivation of $A \vdash \Box A$ but not of $\vdash A \supset \Box A$; thus it is said that 'the deduction theorem fails in modal logic'. When a traditional Hilbert-type system of axiomatic logic is generalized into a system for derivations from assumptions, the necessitation rule has to be modified in a way that restricts its use to cases in which the premiss does not depend on assumptions. This restriction is entirely analogous to the restriction of the rule of universal generalization of first-order logic. A detailed discussion of the issue, together with a proof of the deduction theorem both directly in a Hilbert system extended with assumptions, and indirectly, through equivalence with a cut-free sequent system for basic modal logic, is presented in Hakli and Negri (2011a).

The closure condition (sec. 11.4) could seem to bring back, in some cases, a contraction on relational atoms. However, a simple argument shows that whenever a rule that arises from the closure condition is an instance of contraction (such as *Trans**) it is in fact admissible and need not be added (cf. proposition 3 in Hakli and Negri 2011b).

The development of sequent systems for non-classical logics, in particular modal logics, started in the 1950s with the work of Curry (1952), who provided a system with cut elimination and a decision procedure for **S4**, and Kanger (1957), who gave sequent calculi and decision procedures for **T**, **S4**, and **S5** with the use of 'spotted formulas', formulas indexed by natural numbers.

Difficulties in the Gentzen-style formalization of modal logic were, however, encountered at a very elementary level, for instance in the search for an adequate cut-free sequent calculus for the modal logic **S5**. In 1957 Ohnishi and Matsumoto presented sequent calculi with cut elimination for various modal logics, but no cut elimination for **S5**. Mints (1970) gives a sequent calculus for **S5** with quantifiers that enjoys cut elimination but not the subformula property. The same limitation is encountered in Sato (1980). Shvarts (1989) gave an indirect proof of cut elimination, by showing that A is provable in **S5** if and only if $\Box A$ is provable in a suitable cut-free calculus. A similar idea, translated in terms of tableaux systems, is exploited in Fitting (1999). Braüner (2000) proved cut elimination for a calculus for **S5** that cannot be appropriately called a sequent system, because of the non-locality of its rules. Two recent proposals of a sequent calculus for **S5** appear in Restall (2008) and in Stouppa (2007).

The lack of a general solution has caused an overall pessimistic attitude towards the possibility of applying Gentzen's systems to non-classical logics (Fitting 1983, p. 4, Bull and Segerberg 1984, p. 7, Sally Popkorn 1994, p. 97). Also, in a recent textbook on modal logic, the development of a proof theory for modal logic is presented as a premature undertaking (Blackburn, de Rijke, and Venema 2001, p. xvi).

The failure of ordinary sequent systems is thus seen as a reason for investigating generalizations of the notion of a Gentzen sequent. These generalizations include systems such as higher-level sequents, higher-dimensional, higher-arity, multiple-sequent systems, hypersequents, and display logic (Wansing 1996, 2002).

In addition to these generalizations, in recent years an approach based on the internalization of the Kripke semantics within a calculus has gained prominence. This idea, with early precursors as far back as in Kanger (1957), has been developed in several forms. Inference systems have been presented that incorporate possible worlds in the form of sequents (Mints 1997, Viganò 2000, Kushida and Okada 2003, Castellini and Smaill 2002, Castellini 2005), in the form of tableaux (Fitting 1983, Catach 1991, Nerode 1991, Goré

1998, Massacci 2000), and in the form of natural deduction (Fitch 1966, Simpson 1994, Basin, Matthews, and Viganó 1998). The use of a syntax that includes the relational semantics has been central also in the work on first-order encodings of modal logic (Ohlbach 1993, Schmidt and Hustadt 2003) and in what is called hybrid logic (Blackburn 2000). Internalization of the algebraic – rather than relational – semantics in a natural deduction style presentation is instead mainly used in labelled deductive systems (Gabbay 1996).

The completeness proof of Section 11.7 comes from Negri (2009). This article analyses Kripke's original proofs from 1959 and 1963 and presents a historical reconstruction of the emergence of Henkin-style completeness proofs for modal logic. It also shows how the use of a labelled sequent system permits a direct and uniform completeness proof for a wide variety of modal logics. The proof is close in spirit to Kripke's original arguments, but without the drawbacks of the informal arguments in Kripke's proof or of the implicit character of Henkin-style completeness proofs.

Important references for the history of modal logic and possible worlds semantics are Copeland (2002) and Goldblatt (2005).

12 | Quantified modal logic, provability logic, & other non-classical logics

This chapter begins with the addition of the quantifiers to modal logic. Next provability logic is treated. The last two sections analyse formal proofs in some systems of non-classical logics, namely intermediate logics and substructural logics. The former are logical systems between intuitionistic and classical logic. They are characterized axiomatically by the addition of some axioms that extend intuitionistic logic, such as $(A \supset B) \vee (B \supset A)$. Such axioms often have a semantic equivalent that can be converted into a rule of our labelled calculi. The same goes for the substructural logics in the final section.

12.1 Adding the quantifiers

(a) **Semantics and syntax of quantified modal logic.** We extend our treatment to first-order modal logic through an internalization within the syntax of the notion of a **quantificational model**. In such a model, introduced in Kripke (1963b), there is associated to every world w a domain of interpretation of individual variables $D(w)$. The whole domain of interpretation of individual variables is $D \equiv \bigcup_{w \in K} D(w)$. The valuation of n-ary predicates $P(x_1, \ldots, x_n)$ in a quantificational model is given by a function $[\![P]\!]$ from D^n to the set of classical truth values $\{0, 1\}$, or, in other words, a (classical) subset of D^n. We say that w forces $P(x_1, \ldots, x_n)$ under the **assignment** $\sigma \equiv \langle a_1/x_1, \ldots a_n/x_n \rangle$, written $w \Vdash_\sigma P(x_1, \ldots, x_n)$, if $[\![P]\!](a_1, \ldots, a_n) = 1$. Observe that the notion of quantificational model is not yet fully determined; there is, for example, an ambiguity in the case in which some of the a_i are not in $D(w)$. As we shall see, there are different possible assumptions about the domains $D(w)$ that give rise to different notions of quantificational models. We can nevertheless proceed with a formal analysis of the notion to be imported into our syntactic treatment. The possible distinctions for quantificational models will then be clear. The treatment will also exploit in full the similarities between modalities and quantifiers: modalities are like quantifiers with possible worlds as their scope. The interaction of modalities with the proper quantifiers is captured by conditions that connect possible worlds and ranges of variables.

Given a valuation of atomic predicates under an assignment, it is extended into a valuation of arbitrary formulas in a quantificational model by the standard inductive clauses for propositional connectives. For the quantifiers, we have

$w \Vdash_\sigma \forall x A(x, y_1, \ldots, y_n)$ whenever for all σ' such that $\sigma'(y_i) = \sigma(y_i)$,
$w \Vdash_{\sigma'} \forall x A(x, y_1, \ldots, y_n)$.

Alternatively, by leaving arbitrary the assignment of free variables, we have

$w \Vdash \forall x A(x)$ whenever for all a in $D(w)$, $w \Vdash A(a/x)$.

In a similar way, we have

$w \Vdash \exists x A(x)$ whenever for some a in $D(w)$, $w \Vdash A(a/x)$.

Correspondingly, we add into our calculus expressions of the form $a \in D(w)$ that can appear in sequents together with labelled formulas $w : A$ and relational atoms wRo. The rules for the quantifiers are then obtained directly from the semantic explanations:

Table 12.1 Quantifier rules of **G3Kq**

$$\frac{a \in D(w), \Gamma \to \Delta, w : A(a/x)}{\Gamma \to \Delta, w : \forall x A} \, R\forall$$

$$\frac{w : A(a/x), w : \forall x A, a \in D(w), \Gamma \to \Delta}{w : \forall x A, a \in D(w), \Gamma \to \Delta} \, L\forall$$

$$\frac{a \in D(w), \Gamma \to \Delta, w : \exists x A, w : A(a/x)}{a \in D(w), \Gamma \to \Delta, w : \exists x A} \, R\exists$$

$$\frac{a \in D(w), w : A(a/x), \Gamma \to \Delta}{w : \exists x A, \Gamma \to \Delta} \, L\exists$$

Rules $R\forall$, $L\exists$ have the condition $a \notin \Gamma, \Delta$. To the initial sequents are added sequents of the form $a \in D(w), \Gamma \to \Delta, a \in D(w)$. These sequents are not needed in practice because they, as well as those of the form $wRo, \Gamma \to \Delta, wRo$, are needed only for the derivation of the properties of the accessibility relation and of the domain.

Quantified modal logic is not as intuitive as standard propositional modal logic. For example, the principle of universal instantiation, in the form $\forall x A(x) \supset A(a)$, fails to hold in general. This failure is seen through a proof search by our rules:

$$\vdots$$

$$\frac{w : \forall x A(x) \to w : A(a)}{\to w : \forall x A(x) \supset A(a)} \, R\supset$$

After this single step, no rule is applicable; the only way to continue would be by a step of $L\forall$ but this would require the additional assumption $a \in D(w)$. Once we have proved the completeness of the calculus, it can be concluded that the above failed proof search is exhaustive.

Let **G3Kq** be the system obtained by adding the quantifier rules to **G3K**. Similarly to **G3K**, we may add to **G3Kq** properties of the accessibility relation and obtain, for example, a system for **S5** with quantifiers by adding rules *Ref, Trans,* and *Sym.* However, as we anticipated, there is more that can be done in first-order extensions: in addition to properties of the accessibility relation, also properties of the domain function can be required. For instance, it can be postulated that for every world, the corresponding domain of interpretation be **non-empty**:

$$\forall w \exists a (a \in D(w))$$

Another condition is that domains be **increasing**:

$$\forall wo \forall a (wRo \mathrel{\&} a \in D(w) \supset a \in D(o))$$

They can also be **decreasing**:

$$\forall wo \forall a (wRo \mathrel{\&} a \in D(o) \supset a \in D(w))$$

All the above properties follow the geometric rule scheme and their rule form is as follows, with the variable condition $a \notin \Gamma, \Delta$ in the first:

$$\frac{a \in D(w), \Gamma \to \Delta}{\Gamma \to \Delta} \; Nonempty$$

$$\frac{a \in D(o), wRo, a \in D(w), \Gamma \to \Delta}{wRo, a \in D(w), \Gamma \to \Delta} \; Incr$$

$$\frac{a \in D(w), wRo, a \in D(o), \Gamma \to \Delta}{wRo, a \in D(o), \Gamma \to \Delta} \; Decr$$

The permutability of the necessity modality and the universal quantifier has been the object of a long philosophical discussion (see, e.g., Fitting and Mendelsohn 1998). The **Barcan formula** is $\forall x \Box A \supset \Box \forall x A$ and the **converse Barcan formula** $\Box \forall x A \supset \forall \Box x A$. Each of these has its own implications for the meaning of necessity. Below, we shall give derivations of the Barcan formula and its converse in **G3Kq**+*Decr* and in **G3Kq**+*Incr*, respectively.

The property of non-emptiness is usually part of the ontology of the intended semantics for quantified systems of logic and is implicit in the rule of elimination of the universal quantifier. However, we gain a more flexible approach by not having it inbuilt in the rules for the quantifiers.

The property is formally similar to the property of seriality that was added to **G3K** to obtain deontic logic. This latter property is characterized by the axiom $\Box A \supset \Diamond A$. Similarly, non-emptiness corresponds to the axiom $\forall x A \supset \exists x A$, derived as follows, where the topsequents are derivable by the extension of Lemma 11.2 to **G3Kq**, shown below:

$$\dfrac{\dfrac{\dfrac{\dfrac{w : A(a/x), a \in D(w), w : \forall x A \to w : \exists x A, w : A(a/x)}{w : A(a/x), a \in D(w), w : \forall x A \to w : \exists x A} \, R\exists}{a \in D(w), w : \forall x A \to w : \exists x A} \, L\forall}{w : \forall x A \to w : \exists x A} \, \text{Nonempty}}{\to w : \forall x A \supset \exists x A} \, R\supset$$

(b) Structural properties. Let **G3Kq*** be any extension of **G3Kq** by rules that follow the geometric rule scheme for the atomic predicates $a \in D(w)$, wRo (called, respectively, **domain atoms** and **relational atoms**). Rules that involve domain atoms will be called **domain rules**. We extend the proofs of the structural properties of **G3K** to **G3Kq***. Observe that, unlike derivability, admissibility is not a monotone property of deductive systems, that is, it is not automatically maintained in extensions, so some care is needed. However, because of the modularity of all the extensions considered, it is enough to check, in the inductive proofs, the new cases that arise from the addition of the quantifier rules and of the domain rules.

Lemma 12.1. *Sequents of the form*

$$w : A, \Gamma \to \Delta, w : A$$

with A an arbitrary first-order modal formula, are derivable in **G3Kq***.

Proof. We add the new cases of quantified A to the proof of Lemma 11.2. The inductive steps for $\forall$ and $\exists$ are as follows:

$$\dfrac{\dfrac{w : A(a/x), a \in D(w), w : \forall x A, \Gamma \to \Delta, w : A(a/x)}{a \in D(w), w : \forall x A, \Gamma \to \Delta, w : A(a/x)} \, L\forall}{w : \forall x A, \Gamma \to \Delta, w : \forall x A} \, R\forall$$

$$\dfrac{\dfrac{a \in D(w), w : A(a/x), \Gamma \to \Delta, w : \exists x A, w : A(a/x)}{a \in D(w), w : A(a/x), \Gamma \to \Delta, w : \exists x A} \, R\exists}{w : \exists x A, \Gamma \to \Delta, w : \exists x A} \, L\exists$$

QED.

Theorem 12.2. *The Barcan formula and the converse Barcan formula are derivable in* **G3Kq**$+$ *Decr and in* **G3Kq**$+$ *Incr, respectively.*

Proof. The derivations are as follow, where the topsequents are derivable by Lemma 12.1.

$$
\cfrac{
\cfrac{
\cfrac{
\cfrac{
\cfrac{
\cfrac{o : A(a/x), w : \Box A(a/x), a \in D(w), a \in D(o), wRo, w : \forall x \Box A \to o : A(a/x)}
{w : \Box A(a/x), a \in D(w), a \in D(o), wRo, w : \forall x \Box A \to o : A(a/x)} \; L\Box}
{a \in D(w), a \in D(o), wRo, w : \forall x \Box A \to o : A(a/x)} \; L\forall}
{a \in D(o), wRo, w : \forall x \Box A \to o : A(a/x)} \; Decr}
{wRo, w : \forall x \Box A \to o : \forall x A} \; R\forall}
{w : \forall x \Box A \to w : \Box \forall x A} \; R\Box}
{\to w : \forall x \Box A \supset \Box \forall x A} \; R\supset
$$

$$
\cfrac{
\cfrac{
\cfrac{
\cfrac{
\cfrac{
\cfrac{o : A(a/x), o : \forall x A, a \in D(o), wRo, a \in D(w), w : \Box \forall x A \to o : A(a/x)}
{o : \forall x A, a \in D(o), wRo, a \in D(w), w : \Box \forall x A \to o : A(a/x)} \; L\forall}
{a \in D(o), wRo, a \in D(w), w : \Box \forall x A \to o : A(a/x)} \; L\Box}
{wRo, a \in D(w), w : \Box \forall x A \to o : A(a/x)} \; Incr}
{a \in D(w), w : \Box \forall x A \to w : \Box A(a/x)} \; R\Box}
{w : \Box \forall x A \to \forall x \Box A} \; R\forall}
{\to w : \Box \forall x A \supset \forall x \Box A} \; R\supset
$$

QED.

Lemma 12.3. *For arbitrary A and B, the sequent*

$$\to w : \Box(A \supset B) \supset (\Box A \supset \Box B)$$

is derivable in **G3Kq***.

Proof. By steps of rules for $\Box$ and $\supset$ and Lemma 12.1.   QED.

To deal with **G3Kq***, we need to extend the definition of substitution to domain atoms, as follows:

$$a \in D(w)(b/a) \equiv b \in D(w)$$
$$a \in D(w)(o/w) \equiv a \in D(o)$$

We have:

Lemma 12.4.

(i) *If* $\Gamma \to \Delta$ *is derivable in* **G3Kq***, then also* $\Gamma(o/w) \to \Delta(o/w)$ *is derivable, with the same derivation height.*

(ii) *If* $\Gamma \to \Delta$ *is derivable in* **G3Kq***, then also* $\Gamma(b/a) \to \Delta(b/a)$ *is derivable, with the same derivation height.*

Proof. The proof is by induction on the height n of the derivation of $\Gamma \to \Delta$ and extension of the one for Lemma 11.4. Only the new cases that arise from the addition of the rules for domain atoms and for quantifiers need be considered.

(i) If $n = 0$, we have an initial sequent and the claim holds because the result of substitution is an initial sequent of the same form.

   If $n > 0$ and the last rule in the derivation is $R\forall$ with principal formula labelled by w, we have

$$\frac{\vdots}{a \in D(w), \Gamma \to \Delta', w : A(a/x)}{\Gamma \to \Delta', w : \forall x A} \, R\forall$$

Here $a \notin \Gamma \to \Delta'$. By the inductive hypothesis and application of $R\forall$, we obtain in n steps the following derivation:

$$\frac{\vdots}{a \in D(o), \Gamma(o/w) \to \Delta'(o/w), o : A(a/x)}{\Gamma(o/w) \to \Delta'(o/w), o : \forall x A} \, R\forall$$

Also the other rules for quantifiers have no variable conditions on worlds, so the proof is entirely similar. First the inductive hypothesis is applied to the premiss of the rule, then the rule. The same is true for the domain rules *Nonempty*, *Incr*, and *Decr*.

(ii) For height $n = 0$ the proof is clear. If $n > 0$ and the last rule is a quantifier rule without a variable condition ($L\forall$ or $R\exists$), proceed as detailed above for substitution on worlds. If the last rule is a quantifier rule with a variable condition, then either the substitution is vacuous or a is not a domain eigenvariable of the rule. In the former case there is nothing to prove. In the latter, assume that neither b is an eigenvariable. If the last rule is $R\forall$ we have:

$$\frac{\vdots}{c \in D(w), \Gamma \to \Delta', w : A(c/x)}{\Gamma \to \Delta', w : \forall x A} \, R\forall$$

Because $A(c/x)(b/a) = A(b/a)(c/x)$, this is transformed by the inductive hypothesis and $R\forall$ into:

$$\frac{\vdots}{c \in D(w), \Gamma(b/a) \to \Delta'(b/a), w : A(c/x)(b/a)}{\Gamma(b/a) \to \Delta'(b/a), w : \forall x A(b/a)} \, R\forall$$

If b is a domain eigenvariable, we have

$$\frac{\vdots}{b \in D(w), \Gamma \to \Delta', w : A(b/x)}{\Gamma \to \Delta', w : \forall x A} \, R\forall$$

First apply the inductive hypothesis to replace b with a fresh domain eigenvariable c. By the variable condition, the substitution does not affect Γ and Δ, so we have a derivation of height $n-1$ of $c \in D(w)$, $\Gamma \rightarrow \Delta', w : A(c/x)$. We have, again by the inductive hypothesis, a derivation of height $n-1$ of $c \in D(w), \Gamma(b/a) \rightarrow \Delta'(b/a)$, $w : A(c/x)(b/a)$, hence by $R\forall$ and the identity of $A(c/x)(b/a)$ and $A(b/a)(c/x)$, a derivation of height n of $\Gamma(b/a) \rightarrow \Delta'(b/a)$, $w : \forall x A(b/a)$.

If the last rule in the derivation is a rule for domains, such as *Incl*, without variable conditions, apply the inductive hypothesis to the premiss and then the rule. If it is a rule with variable conditions, such as *Nonempty*, then again either the substitution is vacuous or a is not a variable with conditions. If neither b is a variable with conditions, we have the derivation

$$\frac{\vdots}{\dfrac{c \in D(w), \Gamma \rightarrow \Delta}{\Gamma \rightarrow \Delta}} \; Nonempty$$

Here $c \notin \Gamma, \Delta, c \neq a, c \neq b$, and the derivation is transformed into

$$\frac{c \in D(w), \Gamma(b/a) \rightarrow \Delta(b/a)}{\Gamma(b/a) \rightarrow \Delta(b/a)} \; Nonempty$$

If instead b is the existential variable, we make first a substitution of b by a fresh variable c and then proceed as above.   QED.

For quantified modal logic we have, in addition to the weakening and contraction rules of **G3K**, also weakening and contraction rules that operate on domain atoms. All such rules are admissible, by the following proposition.

Theorem 12.5. *The rules of weakening*

$$\frac{\Gamma \rightarrow \Delta}{w : A, \Gamma \rightarrow \Delta} \; LW \qquad \frac{\Gamma \rightarrow \Delta}{\Gamma \rightarrow \Delta, w : A} \; RW$$

$$\frac{\Gamma \rightarrow \Delta}{wRo, \Gamma \rightarrow \Delta} \; LW_R \qquad \frac{\Gamma \rightarrow \Delta}{\Gamma \rightarrow \Delta, wRo} \; RW_R$$

$$\frac{\Gamma \rightarrow \Delta}{x \in D(w), \Gamma \rightarrow \Delta} \; LW_D \qquad \frac{\Gamma \rightarrow \Delta}{\Gamma \rightarrow \Delta, x \in D(w)} \; RW_D$$

are height-preserving admissible in **G3K***.

Proof. Extend the induction detailed in the proof of Theorem 11.5, observing that if the last step is a quantifier rule with a variable condition, the substitution lemma is applied to the premisses of the rule to have a fresh eigenvariable that does not clash with that in $a \in D(w)$. The conclusion is then obtained by an application of the inductive hypothesis and the quantifier rule. The same procedure is followed if the last step is a geometric rule and $w : A$ or wRo or $a \in D(w)$ contains an eigenvariable.   QED.

Observe that rules RW_R and RW_D are not needed if we use the equivalent calculus that does not contain the initial sequents $wRo, \Gamma \rightarrow \Delta, wRo$ and $a \in D(w), \Gamma \rightarrow \Delta, a \in D(w)$.

Corollary 12.6. *The rule of necessitation is admissible in* **G3Kq***.

Proof. The same as the proof of Corollary 11.6, but using admissibility of weakening and substitution for **G3Kq***.   QED.

Also invertibility is maintained in the extensions of first-order basic modal logic:

Theorem 12.7. *All the rules of* **G3Kq*** *are height-preserving invertible.*

Proof. Extend the proof of Lemma 11.7 by the following cases. Rules $L\forall$ and $R\exists$ are height-preserving invertible because of the height-preserving invertibility of weakening. The same holds for the domain rules.

Height-preserving invertibility of $R\forall$ and $L\exists$ is shown in a way completely similar to the height-preserving invertibility of $R\Box$ and $L\Diamond$, respectively, in Lemma 11.7. Also observe that the invertibilities of Lemma 11.7 are maintained when the quantifier and the domain rules are added.   QED.

In addition to the contraction rules of **G3K**, there are contraction rules to be considered that act on domain atoms:

$$\frac{a \in D(w), a \in D(w), \Gamma \rightarrow \Delta}{a \in D(w), \Gamma \rightarrow \Delta} \, L\text{-}Ctr_D \qquad \frac{\Gamma \rightarrow \Delta, a \in D(w), a \in D(w)}{\Gamma \rightarrow \Delta, a \in D(w)} \, R\text{-}Ctr_D$$

Rule $R\text{-}Ctr_D$ is not needed if we use the calculus without the initial sequents $a \in D(w), \Gamma \rightarrow \Delta, a \in D(w)$.

Theorem 12.8. *The rules of contraction are height-preserving admissible in* **G3Kq***.

Proof. The proof is obtained by adding to the proof of Theorem 11.8 the cases in which the last rule in the derivation is a quantifier rule or a domain rule. Observe that as for the rules for $\Box$ and $\Diamond$, the rules for the quantifiers are either invertible ($R\forall$ and $L\exists$) or are made so by copying the principal formulas into their premises ($L\forall$ and $R\exists$). It follows that a step of contraction that acts on their principal formulas is either reduced to a contraction on a smaller formula or to a contraction of lesser height.

The domain rules follow the general pattern of rules that extend sequent calculi, so the reduction of contraction below a domain rule is made as usual.   QED.

Analogously to the cuts on relational atoms for **G3K***, there are in **G3Kq*** cuts on domain atoms that should, *a priori*, be considered:

$$\frac{\Gamma \to \Delta, a \in D(w) \quad a \in D(w), \Gamma' \to \Delta'}{\Gamma, \Gamma' \to \Delta, \Delta'} \ Cut_D$$

However, these cuts cannot occur in the variant of **G3K** without the initial sequents $a \in D(w), \Gamma \to \Delta, a \in D(w)$, because no domain atom can appear on the right-hand side of sequents in derivations.

Theorem 12.9. *The rule of cut is admissible in* **G3K***.

Proof. The proof extends the proof of Theorem 11.9 to the cases of cuts on quantified formulas and domain atoms. We shall consider here in detail only the cases in which the cut formulas are of the form $w : \forall xA$ and $w : \exists xA$ and are principal in both premisses of a cut.

If the cut formula is $w : \forall xA$, the derivation is

$$\frac{\dfrac{a \in D(w), \Gamma \to \Delta, w : A(a/x)}{\Gamma \to \Delta, w : \forall xA} \ R\forall \qquad \dfrac{w : A(b/x), b \in D(w), w : \forall xA, \Gamma' \to \Delta'}{b \in D(w), w : \forall xA, \Gamma' \to \Delta'} \ L\forall}{b \in D(w), \Gamma, \Gamma' \to \Delta, \Delta'} \ Cut$$

It is transformed into

$$\frac{b \in D(w), \Gamma \to \Delta, w : A(b/x) \qquad \dfrac{\Gamma \to \Delta, w : \forall xA \quad w : A(b/x), b \in D(w), w : \forall xA, \Gamma' \to \Delta'}{w : A(b/x), b \in D(w), \Gamma, \Gamma' \to \Delta, \Delta'} \ Cut}{\dfrac{b \in D(w), b \in D(w), \Gamma, \Gamma, \Gamma' \to \Delta, \Delta, \Delta'}{b \in D(w), \Gamma, \Gamma' \to \Delta, \Delta'} \ Ctr^*} \ Cut$$

Here the upper cut is of lesser height and the lower on a smaller cut formula, *Ctr** denotes repeated applications of contraction rules, and the leftmost premiss is obtained by the substitution (b/a) from $a \in D(w), \Gamma \to \Delta, w : A(a/x)$.

If the cut formula is $w : \exists xA$, the derivation is

$$\frac{\dfrac{a \in D(w), \Gamma \to \Delta, w : \exists xA, w : A(a/x)}{a \in D(w), \Gamma \to \Delta, w : \exists xA} \ R\exists \qquad \dfrac{b \in D(w), w : A(b/x), \Gamma' \to \Delta'}{w : \exists xA, \Gamma' \to \Delta'} \ L\exists}{a \in D(w), \Gamma, \Gamma' \to \Delta, \Delta'} \ Cut$$

It is transformed into

$$\frac{\dfrac{a \in D(w), \Gamma \to \Delta, w : \exists xA, w : A(a/x) \quad w : \exists xA, \Gamma' \to \Delta'}{a \in D(w), \Gamma, \Gamma' \to \Delta, \Delta', w : A(a/x)} \ Cut \qquad a \in D(w), w : A(a/x), \Gamma' \to \Delta'}{\dfrac{a \in D(w), a \in D(w), \Gamma, \Gamma', \Gamma' \to \Delta, \Delta', \Delta'}{a \in D(w), \Gamma, \Gamma' \to \Delta, \Delta'} \ Ctr^*} \ Cut$$

Here the upper cut is of lesser height and the lower on a smaller cut formula, and the rightmost premiss is obtained by the substitution (b/a) from the sequent $b \in D(w)$, $w : A(b/x), \Gamma' \rightarrow \Delta'$. QED.

(c) Soundness and completeness. We shall now define formally the semantics for first-order modal logic that we have already introduced for the justification of the rules for the quantifiers:

Definition 12.10. *A* **quantificational frame** $(K, R, \mathcal{D})$ *is a frame* (K, R) *endowed with sets (domains)* $\mathcal{D}(k)$ *for every element* k *of* K. *Let* X *be the set of variables and* D *the set of labels for elements of the domains in the language of* **G3Kq***. An* **interpretation** *of* D *in* $\mathcal{D}$ *is a function* $[\![\cdot]\!] : D \rightarrow \mathcal{D}$ *and an* **assignment** *of variables is a function* $\sigma : X \rightarrow D$. *A* **valuation** *at world* k *of atomic n-ary predicates* $P(x_1, \ldots, x_n)$ *is a map* $[\![\cdot]\!]_k : AtPr \rightarrow \mathcal{P}(\mathcal{D}^n(k))$.

We say that k **forces** P **under the assignment** $\sigma \equiv \langle a_1/x_1, \ldots, a_n/x_n \rangle$ **for the given interpretation** if $([\![a_1]\!], \ldots, [\![a_n]\!]) \in [\![P]\!]_k$. This forcing relation is written more compactly as $k \Vdash_\sigma P(x_1, \ldots, x_n)$ or $k \Vdash P(a_1, \ldots, a_n)$.

The assignment of free variables can be left arbitrary, so that $w \Vdash P(x_1, \ldots, x_n)$ is the same as $w \Vdash_\sigma P(x_1, \ldots, x_n)$ for an arbitrary σ. In particular, valuations of atomic predicates are extended to arbitrary formulas by the following clauses for the quantifiers:

$k \Vdash_\sigma \forall x A(x, y_1, \ldots, y_n)$ whenever, for all σ' such that $\sigma'(y_i) = \sigma(y_i)$ for $i = 1, \ldots, n$, $k \Vdash_{\sigma'} A(x, y_1, \ldots, y_n)$.

$k \Vdash_\sigma \exists x A(x, y_1, \ldots, y_n)$ whenever for some σ' such that $\sigma'(y_i) = \sigma(y_i)$ for $i = 1, \ldots, n$, $k \Vdash_{\sigma'} A(x, y_1, \ldots, y_n)$.

By letting arbitrary the assignment of free variables, we can restate the inductive clauses for the valuation of arbitrary quantificational formulas:

Definition 12.11. *The valuation of arbitrary formulas in a quantificational model is defined inductively by the propositional clauses of Definition 11.25 together with the following clauses for the quantifiers:*

$k \Vdash \forall x A(x)$ whenever for all $a \in \mathcal{D}(k)$, $k \Vdash A(a/x)$,

$k \Vdash \exists x A(x)$ whenever for some $a \in \mathcal{D}(k)$, $k \Vdash A(a/x)$.

The definition of validity of sequents in the first-order language is as follows:

Definition 12.12. *A sequent is valid if it is valid for every interpretation of labels, every assignment of variables, and every valuation of atomic predicates in a quantificational frame.*

Theorem 12.13. Soundness. *If sequent $\Gamma \to \Delta$ is derivable in* **G3Kq***, then it is valid in every frame with the properties* *.

Proof. By induction on the height of the derivation. By the proof of Theorem 11.23, we have to consider only the new cases that arise from the addition of the quantifiers and of the domain rules.

Let the last rule in the derivation be $R\forall$ with premiss $a \in D(w)$, $\Gamma' \to \Delta'$, $w : A(a/x)$ and assume that the premiss is valid. Let $[[\cdot]]$, σ be an arbitrary interpretation and an arbitrary assignment that validates all the formulas in Γ'. We claim that one of the formulas in Δ' or $w : \forall A$ is valid under this intepretation and assignment. Let a be an arbitrary element of $D(w)$ and let σ' be the assignment identical to σ except possibly on x, where we set $\sigma(x) = a$. This assignment and interpretation validates all the formulas in the antecedent of the premiss, so it validates a formula in Δ' or $w : A(a/x)$. In the former case we have the result that a formula in Δ' is validated also under the assignment σ, in the latter that $w : \forall xA$ is valid.

If the last rule in the derivation is $L\forall$, with premiss $w : A(a/x)$, $a \in D(w)$, $w : \forall xA$, $\Gamma' \to \Delta$, assume an arbitrary valuation and assignment that validates $a \in D(w)$, $w : \forall xA$, Γ'. By the definition of validity of a universal formula, also $A(a/x)$ is valid at w, and therefore the validity of the premiss of the rule gives the desired conclusion.

Validity of the domain rules is proved in the same way as validity of the rules for relational atoms; for instance, rule *Incr* is valid in all frames with increasing domains, and rule *Nonempty* is valid in all frames with non-empty domains.   QED.

Theorem 12.14. Completeness. *Let $\Gamma \to \Delta$ be a sequent in the language of* **G3Kq***. Then either the sequent is derivable in* **G3Kq** *or it has a Kripke countermodel with properties* *.

Proof. The reduction tree is built as in the proof of Theorem 11.24, with the following additional stages for the quantifier and domain rules:

For $n = 11$, we consider all topsequents with quantificational formulas $w_1 : \forall x_1 A_1, \ldots, w_m : \forall x_m A_m$ and the domain atoms $a_1 \in D(w_1)$, $\ldots$, $a_m \in D(w_m)$ in the antecedent, and write on top of these the sequent $w_1 : A_1(a_1/x_1), \ldots, w_m : A_1(a_m/x_m), a_1 \in D(w_1), \ldots, a_m \in D(w_m)$, $w_1 : \forall x_1 A_1, \ldots, w_m : \forall x_m A_m, \Gamma' \to \Delta$. This step corresponds to m applications of rule $L\forall$.

For $n = 12$, let $w_1 : \forall x_1 A_1, \ldots, w_m : \forall x_m A_m$ be all the formulas with $\forall$ as outermost connective in the succedent of topsequents of the tree, and let Δ' be the other formulas. Let $b_1, \ldots, b_m$ be fresh labels for domain elements, not yet used in the reduction tree, and write on top of each sequent the sequent

$$b_1 \in D(w_1), \ldots, b_m \in D(w_m), \Gamma \to \Delta, w_1 : A_1(b_1/x_1), \ldots,$$

$$w_m : A_1(b_m/x_m)$$

This step corresponds to m applications of rule $R\forall$.

For $n = 13$, let $w_1 : \exists x_1 A_1, \ldots, w_m : \exists x_m A_m$ be all the formulas with $\exists$ as outermost operation in the antecedent of the topsequents of the tree, and let Γ' be the other formulas. Let $b_1, \ldots, b_m$ be fresh labels, and write on top of each sequent the sequent

$$b_1 \in D(w_1), \ldots, b_m \in D(w_m), w_1 : A_1(b_1/x_1), \ldots,$$

$$w_m : A(b_m/x_m), \Gamma' \to \Delta$$

This step corresponds to m applications of rule $L\exists$.

For $n = 14$, consider all topsequents with the formulas $w_1 : \exists x_1 A_1$, $\ldots, w_m : \exists x_m A_m$ in the succedent and domain atoms $a_1 \in D(w_1)$, $\ldots, a_m \in D(w_m)$ in the antecedent, and write on top of these sequents the sequents

$$a_1 \in D(w_1), \ldots, a_m \in D(w_m), \Gamma \to \Delta',$$

$$w_1 : \exists x_1 A_1, \ldots, w_m : \exists x_m A_m,$$

$$w_1 : A_1(a_1/x_1), \ldots, w_m : A(a_m/x_m)$$

This step corresponds to m applications of rule $R\exists$.

For $n = 14 + j$, apply the rules for relational and domain atoms whenever the sequent matches the conclusion of one of them: for instance, if the system includes *Incr* and the sequent under examination contains wRo, $a \in D(w)$ in the antecedent, write on top of this the sequent with $a \in D(o)$ added to the antecedent.

As for the propositional case, we either get a proof from the reduction tree, or, if the search does not terminate, an infinite branch that is used for the construction of the countermodel. The quantificational frame is defined by starting from the propositional frame, with domains $D(w)$ given by all the $a \in D(w)$ in Γ. The relational and the domain rules impose the corresponding frame properties on the countermodel. The valuation of atomic predicate formulas is defined by positing $w \Vdash P(x_1, \ldots, x_n)$ for $P(x_1, \ldots, x_n)$ in Γ and $w \nVdash Q(x_1, \ldots, x_n)$ for $Q(x_1, \ldots, x_n)$ in Δ.

To show that all the formulas in Γ and no formula in Δ are forced by the countermodel, we supplement the proof of Theorem 11.24 with the cases of quantificational formulas. If $w : \forall x A$ is in Γ, consider all the domain atoms $a \in D(w)$. If there is no such atom, then by definition, $\forall x A$ is forced at w in the model. Or else we find $w : A(a/x)$ in Γ, forced in the model by the inductive hypothesis, and therefore $w : \forall x A$ is forced as well. If $w : \forall x A$ is in Δ, we find by construction $a \in D(w)$ in Γ and $w : A(a/x)$ in Δ. By the inductive hypothesis, $A(a/x)$ is not forced at w in the model, where a is an element of the domain $D(w)$; thus $w : \forall x A$ is not forced either.

The cases with $w : \exists x A$ in Γ and $w : \exists x A$ in Δ are symmetric to those of $w : \forall x A$ in Δ and $w : \forall x A$ in Γ, respectively.   QED.

12.2 Provability logic

Gödel–Löb provability logic, nowadays commonly called **GL**, is the logic of arithmetic provability, with $\Box A$ interpreted as 'A is provable in Peano arithmetic'. **GL** can be characterized axiomatically as follows. Start from the system of basic modal logic **K** and add to its axioms and rules, as in Section 11.1, the following **Löb axiom**:

$$\Box(\Box A \supset A) \supset \Box A.$$

The axiom contains the unprovability of consistency as a special case. It can be read as: *if it is provable that from the provability of A, A follows, then A is provable.* In particular, with $\bot$ in place of A, $\Box\bot \supset \bot$ states that falsity is not provable, i.e., that the system of proof in question is consistent. By Löb's axiom, if that is provable, falsity is provable.

After Solovay's landmark paper (1976) that presented **GL** axiomatically as the logic of arithmetic provability and characterized its Kripke models as the transitive and Noetherian frames, a lot of interest has been directed to the search for an adequate, cut-free sequent system for **GL**. The issue is still an open research interest (see the discussion in the notes to this chapter).

In the Kripke frames for provability logic, the accessibility relation R is irreflexive, transitive, and Noetherian, the last condition meaning that every R-chain eventually becomes stationary. Equivalently, we can say that R is transitive and all R-chains are finite. Clearly, this characterizing frame condition is not first order, so the method of universal and geometric extensions exploited in Chapter 11 cannot be applied directly. However, the condition can be internalized in the meaning explanation of the modality as follows:

Lemma 12.15. *In irreflexive, transitive, and Noetherian Kripke frames,*

$$w \Vdash \Box A \text{ whenever for all } o, \text{ from } wRo \text{ and } o \Vdash \Box A \text{ follows } o \Vdash A.$$

Proof. Assume $w \Vdash \Box A$ and let o be such that wRo. Then $o \Vdash A$ holds and the assumption $o \Vdash \Box A$ is superfluous.

For the converse, assume the right-hand side and suppose that $w \nVdash \Box A$. Then there exists w_1 such that wRw_1 and $w_1 \nVdash A$. From the assumption it follows that $w_1 \nVdash \Box A$; hence there exists some w_2 such that $w_1 R w_2$ and $w_2 \nVdash A$. By transitivity, we have wRw_2 and so from the assumption, $w_2 \nVdash \Box A$ follows. In this way, we build a chain $wRw_1, w_1 Rw_2, \ldots$ that never becomes stationary, because of irreflexivity, so we have a contradiction.   QED.

The right-to-left direction of the implication stated above gives the right rule for $\Box$:

$$\frac{wRo, o : \Box A, \Gamma \to \Delta, o : A}{\Gamma \to \Delta, w : \Box A} \; R\Box\text{-}L$$

The variable condition is that o is not in the conclusion. The left-to-right direction gives the left rule:

$$\frac{w : \Box A, wRo, \Gamma \to \Delta, o : \Box A \quad o : A, w : \Box A, wRo, \Gamma \to \Delta}{w : \Box A, wRo, \Gamma \to \Delta} \; L\Box\text{-}L$$

Irreflexivity is the zero-premiss rule that lets us derive $wRw, \Gamma \to \Delta$. The system **G3GL** is thus determined:

Table 12.2 The sequent calculus **G3GL**

Initial sequents

$\quad\quad w : P, \Gamma \to \Delta, w : P \quad\quad w : \Box A, \Gamma \to \Delta, w : \Box A$

Logical rules

$\quad\quad$ As in **G3K** for $\&, \vee, \supset, \bot$; $L\Box$-L, $R\Box$-L

Mathematical rules

$\quad\quad$ *Irref, Trans*

It is an easy task to verify that all the preliminary results, from Lemma 11.2 to Corollary 11.6, proved for **G3K***, continue to hold for **G3GL**. In particular, we have:

Theorem 12.16. *The rules of substitution, weakening, and necessitation are height-preserving admissible in* **G3GL***.

In addition, we have the invertibility, not necessarily height preserving, of all the rules of **G3GL** and of contraction:

Theorem 12.17. *All the rules of the system* **G3GL** *are invertible.*

Proof. For the invertibility of the rules for &, $\vee$, $\supset$, see Lemma 11.7.

Rule $L\square$-L is height-preserving invertible by height-preserving invertibility of weakening. We show invertibility of $R\square$-L by induction on the height n of the derivation of $\Gamma \rightarrow \Delta, w : \square A$. If $n = 0$ and $w : \square A$ is not principal, then also $wRo, o : \square A, \Gamma \rightarrow \Delta, o : A$ is an initial sequent, or an instance of *Irref.* If it is principal, we have $\Gamma \equiv w : \square A, \Gamma'$, and we need to prove $wRo, o : \square A, w : \square A, \Gamma' \rightarrow \Delta, o : A$ derivable. This follows by $L\square$-L from the initial sequent $wRo, o : \square A, w : \square A, \Gamma' \rightarrow \Delta, o : A$, $o : \square A$ and the derivable sequent $o : A, wRo, o : \square A, w : \square A, \Gamma' \rightarrow \Delta$, $o : A$. The inductive step is dealt with as for **G3K**.   QED.

We shall assume, without loss of generality, that derivations are **pure**, i.e., that the eigenvariables used at steps of $R\square$-L appear only in the subtree above the rule that introduces them. Clearly, by height-preserving substitution, such a condition can always be met.

Before proving the admissibility of contraction, we introduce the notion of the **range** of a world in a derivation, to be used as one inductive parameter in the proof of cut elimination. Roughly, the range of a label w in a derivation $\mathcal{D}$ is the set of labels accessible through a chain from w in the left-hand side of sequents of $\mathcal{D}$. It is defined formally as follows:

Definition 12.18. *The range of w in a derivation $\mathcal{D}$ is the (finite) set of worlds o such that either wRo or for some $n \geqslant 1$ and for some $w_1, \ldots, w_n$, each of the atoms $wRw_1, w_1Rw_2, \ldots, w_nRo$ appear in the antecedent of a sequent in $\mathcal{D}$. Ranges of variables are ordered by set inclusion.*

We say that a rule is **range-preserving** admissible if the elimination of the rule does not increase the ranges of variables in the derivation.

Theorem 12.19. *The rules of contraction are range-preserving admissible in* **G3GL**.

Proof. By simultaneous induction for left and right contraction, with induction on the size of the contraction formula and subinduction on derivation height. We detail the proof in one case specific to **G3GL**. Assume we have proved admissibility of contraction for formulas of a size up to n on the left and up to $n - 1$ on the right and assume the contraction formula is $w : \square A$ on the right, of size n. If the last rule in the derivation is not $R\square$-L on the contraction formula, we apply the inductive hypothesis to the premiss of the rule (of smaller height) and then apply the rule. If the last step is $R\square$-L, the premiss is $wRo, o : \square A, \Gamma \rightarrow \Delta, w : \square A, o : A$. By using invertibility of $R\square$-L, we derive a sequent of the form $wRo, wRo, o : \square A, o : \square A$,

$\Gamma \to \Delta, o : A, o : A$, and by using the inductive hypotheses we obtain a derivation of $wRo, o : \Box A, \Gamma \to \Delta, o : A$, hence the conclusion of contraction by $R\Box\text{-}L$.

Although invertibility of $R\Box\text{-}L$ is not, in general, range-preserving, because it introduces a new world, the special instance of invertibility used here does not, as the world needed in the inversion is already a label used in the derivation. It follows that contraction is range-preserving admissible. QED.

The need for the notion of range becomes clear from the proof of cut elimination for **G3GL**.

A typical procedure of cut elimination for **G3**-like systems considers topmost cuts and contains reductions that either decrease the height of one of the two premisses of cut (for permutation cuts, that is, cuts in which the cut formula is not principal in at least one of the premisses) or the size of the cut formula (for detour, or principal, cuts, that is, cuts in which the formula is principal in both premisses). The reductions are repeated until cuts reach initial sequents and disappear. This procedure does not work for **G3GL** in the case of detour cuts on $w : \Box A$. Consider a principal cut on $w : \Box A$:

$$
\cfrac{\cfrac{wRo, o : \Box A, \Gamma \to \Delta, o : A}{\Gamma \to \Delta, w : \Box A} \; R\Box\text{-}L \qquad \cfrac{wRr, w : \Box A, \Gamma' \to \Delta', r : \Box A \quad r : A, wRr, w : \Box A, \Gamma' \to \Delta'}{wRr, w : \Box A, \Gamma' \to \Delta'} \; L\Box\text{-}L}{wRr, \Gamma', \Gamma \to \Delta, \Delta'} \; Cut
$$

This derivation is transformed into four cuts as follows:

$$
\cfrac{\cfrac{\begin{matrix}\mathcal{D}_1 \\ \vdots \\ wRr, wRr, \Gamma', \Gamma, \Gamma \to \Delta, \Delta, \Delta', r : A\end{matrix} \qquad \begin{matrix}\mathcal{D}_2 \\ \vdots \\ wRr, r : A, \Gamma', \Gamma \to \Delta, \Delta'\end{matrix}}{wRr, wRr, wRr, \Gamma', \Gamma', \Gamma, \Gamma, \Gamma \to \Delta, \Delta, \Delta, \Delta', \Delta'} \; Cut}{wRr, \Gamma', \Gamma \to \Delta, \Delta'} \; Ctr^*
$$

Here $\mathcal{D}_1$ and $\mathcal{D}_2$ are the following two derivations:

$$
\cfrac{\cfrac{\Gamma \to \Delta, w : \Box A \quad wRr, w : \Box A, \Gamma', \to \Delta', r : \Box A}{wRr, \Gamma', \Gamma \to \Delta, \Delta', r : \Box A} \; Cut \qquad wRr, r : \Box A, \Gamma \to \Delta, r : A}{wRr, wRr, \Gamma, \Gamma', \Gamma \to \Delta, \Delta', \Delta, r : A} \; Cut
$$

$$
\cfrac{\Gamma \to \Delta, w : \Box A \quad wRr, w : \Box A, r : A, \Gamma' \to \Delta'}{wRr, r : A, \Gamma', \Gamma \to \Delta, \Delta'} \; Cut
$$

Observe that the cuts on $w : \Box A$ and on $r : A$ are all reduced according to the standard inductive parameter, whereas the cut on $r : \Box A$ is not, because neither the complexity of the cut formula nor the height of the cut is reduced.

However, if the range of r in the new derivation is strictly smaller than the range of w in the original derivation, then we have for all the cuts in the transformed derivation a reduced inductive parameter given by the triple that consists of the complexity of the cut formula, the range of its label, and the height of the cut, ordered lexicographically.

To prove the reduction in range, two extra assumptions are needed, namely that there be no cuts such that wRw or $wRw_1, \dots, w_nRw$ appears in the antecedents of their conclusions, and that eigenvariables be pure, i.e., that they appear only in the subtree above the step that introduces them. The first condition is met by observing that if there are cuts of the stated form, they are eliminated by *Irref* and *Trans*. The second condition is met by a fresh renaming of eigenvariables. It then follows that no w can be in the range of itself, that if o is in the range of w, then the range of o is properly included in the range of w, and that if o, r are in the range of w and o is an eigenvariable, then the union of the range of o and the range of r is properly included in the range of w.

The proof of cut elimination for **G3GL** is structured as the proof for **G3K***, but with the induction on weight and height replaced by an induction on the following triple parameter ordered lexicographically:

1. Size of the cut formula,
2. Range of the label of the cut formula,
3. Sum of the heights of the derivations of the two premisses of the cut.

The cases in which the cut formula is not principal in both premisses of cut are dealt with as usual, with the additional observation that permutations do not increase the range since they change neither the cut formula nor its label. The only case specific to **G3GL** has been detailed above. We have thus proved:

Theorem 12.20. *The rule of cut is admissible in* **G3GL**.

As an application of the cut-free calculus we have:

Corollary 12.21. Second incompleteness theorem. *The sequent that expresses consistency,* $\to w : \neg\,\Box\bot$, *is not derivable in* **G3GL**.

Proof. Proceeding root first, if a derivation exists, it ends with

$$\frac{w : \Box\bot \to w : \bot}{\to w : \Box\bot \supset \bot}\; R\supset$$

However, no rule of **G3GL** is applicable to the premiss.   QED.

Finally, in order to obtain the derivability of the characteristic axiom of **GL** we show:

Lemma 12.22. *All sequents of the form $wRo, w : \Box A, \Gamma \to \Delta, o : \Box A$ are derivable in* **G3GL**.

Proof. Root first, by steps of $R\Box\text{-}L$, *Trans*, and $L\Box\text{-}L$.   QED.

Corollary 12.23. *The standard rule $L\Box$ is derivable in* **G3GL**.

Proof. By Lemma 12.22, the left premiss of $L\Box\text{-}L$ is derivable in **G3GL**.   QED.

Even if the two left rules for $\Box$ are interderivable, the use of $L\Box\text{-}L$ seems essential in the proof of cut elimination. If instead the standard rule $L\Box$ were used, a cut with a (derived) sequent of the form $wRo, w : \Box A, \Gamma \to \Delta, o : \Box A$ would be needed. However, its derivation introduces new worlds, thus breaking the property of range admissibility of all cut reductions.

Corollary 12.24. *The Löb axiom is derivable in* **G3GL**.

Proof. Using Corollary 12.23, we have the inference:

$$\cfrac{\cfrac{\cfrac{o : \Box A \supset A, wRo, w : \Box(\Box A \supset A), o : \Box A \to o : A}{wRo, w : \Box(\Box A \supset A), o : \Box A \to o : A} {\scriptstyle L\Box}}{w : \Box(\Box A \supset A) \to w : \Box A} {\scriptstyle R\Box\text{-}L}}{\to w : \Box(\Box A \supset A) \supset \Box A} {\scriptstyle R\supset}$$

The top sequent is derivable by $L\supset$ and Lemma 11.2 for **G3GL**.   QED.

By Corollary 12.23, the system **G3GL**, with rules $R\Box\text{-}L$ and $L\Box\text{-}L$, and the system with rules $R\Box\text{-}L$ and $L\Box$ that we shall call **G3KGL** are equivalent. In the latter system, initial sequents can be restricted to atomic formulas, as in **G3K**, and therefore stronger structural properties such as height-preserving admissibility of contraction hold with no limitations. Cut elimination for **G3KGL** can be established through a translation to **G3GL**, cut elimination in this system, and a translation back to **G3KGL**.

12.3 Intermediate logics

It is well known that intuitionistic logic can be embedded into the classical modal logic **S4**, and actually all the logics intermediate between intuitionistic and classical logic can be embedded in the modal logics intermediate between **S4** and **S5**. The analogy between these two families of logics is best seen at the level of their Kripke semantics. The explanation of the meaning of implication in intuitionistic logic reflects the explanation of the modality in **K**. We can internalize, as we have done for normal modal logics, the inductive definition of validity in a Kripke frame, to obtain uniform **G3**-style sequent calculi for intermediate logics. The accessibility relation for intuitionistic logic is a partial order. The addition of properties to those of reflexivity and transitivity gives logics above intuitionistic logic. We observe that all the properties of the accessibility relation that characterize the **interpolable** propositional logics fall under the geometric rule scheme. By the results on geometric extensions, we obtain for these logics complete calculi with good structural properties. In addition, the uniformity in the syntax allows immediate proofs of the faithfulness of the embeddings.

The inductive definition of validity of an implication in intuitionistic logic in terms of Kripke semantics is:

$$w \Vdash A \supset B \text{ whenever for all } o, \text{ from } w \leqslant o \text{ and } o \Vdash A \text{ follows } o \Vdash B.$$

As before, we can convert this meaning explanation into a system of left and right rules for intuitionistic implication. Arbitrariness of o in the right rule is again expressed by a variable condition.

The rules for the other connectives are exactly as the rules in **G3K**. The initial sequents of **G3K** are instead modified in order to guarantee the property of monotonicity of the forcing relation. In compliance with the features of the **G3**-style calculi, it is enough to have monotonicity with respect to atomic formulas, with full monotonicity admissible. The mathematical rules for the accessibility relation $\leqslant$ are the rules *Ref* and *Trans* that express that $\leqslant$ is a partial order. We have thus determined the following system **G3I** for intuitionistic propositional logic:

Table 12.3 The sequent calculus **G3I**

Initial sequents $w \leqslant o, w : P, \Gamma \rightarrow \Delta, o : P$
Logical rules
As in **G3K** for &, $\vee$, $\bot$

$$\frac{w \leqslant o, w : A \supset B, \Gamma \rightarrow o : A, \Delta, \quad w \leqslant o, w : A \supset B, o : B, \Gamma \rightarrow \Delta}{w \leqslant o, w : A \supset B, \Gamma \rightarrow \Delta} \, L\supset$$

$$\frac{w \leqslant o, o : A, \Gamma \rightarrow \Delta, o : B}{\Gamma \rightarrow \Delta, w : A \supset B} \, R\supset$$

Order rules

$$\frac{w \leqslant w, \Gamma \rightarrow \Delta}{\Gamma \rightarrow \Delta} \, Ref \qquad \frac{w \leqslant r, w \leqslant o, o \leqslant r, \Gamma \rightarrow \Delta}{w \leqslant o, o \leqslant r, \Gamma \rightarrow \Delta} \, Trans$$

Rule $R\supset$ has the condition that o must not be in Γ, Δ.

Let **G3I*** be any extension of **G3I** by rules that follow the geometric rule scheme. The structural properties of **G3I*** are proved uniformly for any extension, after the method presented in Section 11.4.

Lemma 12.25. *All sequents of the following form are derivable in* **G3I**:

(i) $w \leqslant o, w : A, \Gamma \rightarrow \Delta, o : A$

(ii) $w : A, \Gamma \rightarrow \Delta, w : A$

Proof. By mutual induction on the structure of A. The implication from 1 to 2 for all A is routine by *Ref*. For atoms P and for $\perp$, the proof of 1 is trivial. For $A \equiv B \supset C$ we have the following derivation of 1:

$$\cfrac{\cfrac{\ldots, w : B \supset C, r : B, \Gamma \rightarrow \Delta, r : C, r : B \quad \ldots, w : B \supset C, r : C, r : B, \Gamma \rightarrow \Delta, r : C}{\cfrac{\cfrac{w \leqslant o, o \leqslant r, w \leqslant r, x : B \supset C, r : B, \Gamma \rightarrow \Delta, r : C}{w \leqslant o, o \leqslant r, x : B \supset C, r : B, \Gamma \rightarrow \Delta, r : C} \; Trans}{w \leqslant o, w : B \supset C, \Gamma \rightarrow o : B \supset C, \Delta} \; R\supset} \; L\supset}$$

Here the topsequents are derivable by the inductive hypothesis for 2. The cases in which A is a conjunction or disjunction are handled by the inductive hypothesis for 1.   QED.

Substitution of labels in relational atoms of the form $w \leqslant o$ and in labelled formulas $w : A$ follows the definition given in Chaper 11. We have:

Lemma 12.26. *If* $\Gamma \rightarrow \Delta$ *is derivable in* **G3I***, *then* $\Gamma(o/w) \rightarrow \Delta(o/w)$ *is also derivable, with the same derivation height.*

Proof. By induction on the height n of the derivation of $\Gamma \rightarrow \Delta$.

If $n = 0$, and (o/w) is not a vacuous substitution, the sequent can either be an initial sequent of the form $w \leqslant o, w : P, \Gamma' \rightarrow \Delta', o : P$ or of the form $\perp, \Gamma' \rightarrow \Delta$. In each case $\Gamma(o/w) \rightarrow \Delta(o/w)$ is either an initial sequent of the same form or a conclusion of $L \perp$.

Suppose $n > 0$, and consider the last rule applied in the derivation. If it is a rule for & or ∨, apply the inductive hypothesis to the premisses of the rule, and then the rule. Proceed similarly if the last rule is $L\supset$. If the last rule is $R\supset$ and w an eigenvariable of the rule, the substitution is vacuous. Otherwise, if o is not an eigenvariable either, apply the inductive hypothesis to the shorter derivation of the premiss, and then $R\supset$.

If o is the eigenvariable, apply first the inductive hypothesis for the replacement of the eigenvariable o with a fresh variable r. By the variable condition, the substitution does not affect the context, and proceed as in the previous case.

For extensions of **G3I** by geometric rules, some care is needed for avoiding a clash with the eigenvariables of the geometric rule scheme. The details are similar to those in the proof of Lemma 11.4.   QED.

Theorem 12.27. *The rules of weakening,*

$$\frac{\Gamma \to \Delta}{w : A, \Gamma \to \Delta} \, LW \qquad \frac{\Gamma \to \Delta}{\Gamma \to \Delta, w : A} \, RW \qquad \frac{\Gamma \to \Delta}{w \leqslant o, \Gamma \to \Delta} \, LW_{\leqslant}$$

are height-preserving admissible in **G3I***.*

Proof. Straightforward induction on the height of the derivation of the premiss for the rules for &, $\vee$, and $L \supset$. If the last step is $R \supset$, the substitution lemma is applied to the premisses of the rule in order to have a fresh eigenvariable that does not clash with those in $w : A$ or $w \leqslant o$. The conclusion is then obtained by applying the inductive hypothesis and the rule. An identical procedure is applied if the last step is a geometric rule and $w : A$ or $w \leqslant o$ contain some of its eigenvariables.   QED.

To prove height-preserving admissibility of contraction, we need to show the height-preserving invertibility of the rules of the sequent calculi **G3I***.

Theorem 12.28. *All rules of* **G3I*** *are height-preserving invertible.*

Proof. The proof of height-preserving invertibility for the rules for & and $\vee$ is done exactly as for **G3c** (Theorem 6.2). Rule $L \supset$ is height-preserving invertible by Theorem 12.27.

For $R \supset$, we use induction on the height n of the derivation of $\Gamma \to \Delta, w : A \supset B$. If $n = 0$, it is an initial sequent or a conclusion of $L \perp$, but then also $w \leqslant o, o : A, \Gamma \to \Delta, o : B$ is an initial sequent or a conclusion of $L \perp$. Observe that it is essential here that the initial sequents are restricted to atomic formulas.

If $n > 0$ and $\Gamma \to \Delta, w : A \supset B$ is concluded by any rule $\mathcal{R}$ other than $R \supset$, apply the inductive hypothesis to the premiss $\Gamma' \to \Delta', w : A \supset B$, and possibly also to $\Gamma'' \to \Delta'', w : A \supset B$, to obtain derivations of height at most $n - 1$ of the sequent $w \leqslant o, o : A, \Gamma' \to \Delta', o : B$, and possibly also of $w \leqslant o, o : A, \Gamma'' \to \Delta'', o : B$. Application of rule $\mathcal{R}$ gives a derivation of height n of $w \leqslant o, o : A, \Gamma \to \Delta, o : B$. If $\Gamma \to \Delta, w : A \supset B$ is a conclusion of $R \supset$ with principal formula in Δ, proceed in a similar way.

If instead the principal formula is $A \supset B$, the premiss of the last step gives the conclusion, by height-preserving substitution if necessary.   QED.

Theorem 12.29. *The rules of contraction,*

$$\frac{w : A, w : A, \Gamma \to \Delta}{w : A, \Gamma \to \Delta} \, LC \qquad \frac{\Gamma \to \Delta, w : A, w : A}{\Gamma \to \Delta, w : A} \, RC$$

$$\frac{w \leqslant o, w \leqslant o, \Gamma \to \Delta}{w \leqslant o, \Gamma \to \Delta} \, LC_{\leqslant}$$

are height-preserving admissible in **G3I***.

Proof. By simultaneous induction on the derivation height.

If $n = 0$, the premiss is either an initial sequent or a conclusion of $L\bot$. In each case the contracted sequent is also an initial sequent or a conclusion of $L\bot$.

If $n > 0$, consider the last step, by some rule $\mathcal{R}$, used to derive the premiss of the contraction step. If the contraction formula is not principal in $\mathcal{R}$, both occurrences are found in the premisses of the step and have a smaller derivation height. By the induction hypothesis, they can be contracted and the conclusion obtained by applying rule $\mathcal{R}$ to the contracted premisses.

If the contraction formula is principal in rule $\mathcal{R}$, we distinguish three cases: either $\mathcal{R}$ is a rule in which the principal formulas appear also in the premiss, such as $L\supset$ or the rules for $\leqslant$, or it is a rule in which the premisses consist of proper subformulas of the conclusion, such as the rules for & and $\vee$, or it is a rule, in fact $R\supset$, in which the premisses consist of atoms $w \leqslant o$ and proper subformulas of the conclusion. In the first case, contraction is applied, by the induction hypothesis, to the premisses of the rule. If both contraction formulas are principal in a rule for $\leqslant$, the conclusion holds by the closure condition.

In the second case, contraction is reduced to contraction on smaller formulas as in the standard proof for **G3c**.

In the third case, both a subformula of the contraction formula and an atom $w \leqslant o$ are found in the premiss, for instance

$$\frac{w \leqslant o, o : A, \Gamma \to \Delta, o : B, w : A \supset B}{\Gamma \to \Delta, w : A \supset B, w : A \supset B} \, R\supset$$

By the height-preserving invertibility of $R\supset$ applied to the premiss, we obtain a derivation of height at most $n - 1$ of

$$w \leqslant o, w \leqslant o, o : A, o : A, \Gamma \to \Delta, o : B, o : B$$

Now we have, by the induction hypothesis for both forms of contraction, a derivation of height at most $n - 1$ of

$$w \leqslant o, o : A, \Gamma \to \Delta, o : B$$

The conclusion $\Gamma \to \Delta, w : A \supset B$ follows in one more step by $R\supset$.   QED.

Theorem 12.30. *The rule of cut,*

$$\frac{\Gamma \to \Delta, w : A \quad w : A, \Gamma' \to \Delta'}{\Gamma, \Gamma' \to \Delta, \Delta'} \ Cut$$

is admissible in **G3I***.

Proof. The proof has the same structure as the proof of Theorem 11.9. We observe that in all the cases of permutation of cuts that may give a clash with the variable conditions in the implication rules, and in the rules for $\leqslant$ in the case of geometric extensions, an appropriate substitution (Lemma 12.26) prior to the permutation is used.

The proof is thus by induction on the length of the cut formula, with a subinduction on the sum of the heights of the derivations of the premisses of cut. We consider in detail only the case of a cut with the cut formula principal in implication rules in both premisses.

If the cut formula is $w : A \supset B$, the derivation is

$$\frac{\dfrac{w \leqslant r, r : A, \Gamma \to \Delta, r : B}{\Gamma \to \Delta, w : A \supset B} \quad \dfrac{w \leqslant o, w : A \supset B, \Gamma' \to \Delta', o : A \quad w \leqslant o, w : A \supset B, o : B, \Gamma' \to \Delta'}{w \leqslant o, w : A \supset B, \Gamma' \to \Delta'} \ Cut}{w \leqslant o, \Gamma, \Gamma' \to \Delta, \Delta'}$$

It is transformed into

$$\frac{\dfrac{\vdots \qquad\qquad \vdots}{(w \leqslant o)^2, \Gamma^2, \Gamma' \to \Delta^2, \Delta', o : B \quad w \leqslant o, o : B, \Gamma, \Gamma' \to \Delta, \Delta'} \ Cut}{\dfrac{(w \leqslant o)^3, \Gamma^3, \Gamma'^2 \to \Delta^3, \Delta'^2}{w \leqslant o, \Gamma, \Gamma' \to \Delta, \Delta'} \ Ctr^*}$$

The first premiss is derived by

$$\frac{\dfrac{\Gamma \to \Delta, w : A \supset B \quad w \leqslant o, w : A \supset B, \Gamma' \to \Delta', o : A}{w \leqslant o, \Gamma, \Gamma' \to \Delta, \Delta', o : A} \ Cut \quad \dfrac{w \leqslant r, r : A, \Gamma \to \Delta, r : B}{w \leqslant o, o : A, \Gamma \to \Delta, o : B} \ (o/r)}{(w \leqslant o)^2, \Gamma^2, \Gamma' \to \Delta^2, \Delta', o : B} \ Cut$$

The second premiss is derived by

$$\frac{\Gamma \to \Delta, w : A \supset B \quad w \leqslant o, w : A \supset B, o : B, \Gamma' \to \Delta'}{w \leqslant o, o : B, \Gamma, \Gamma' \to \Delta, \Delta'} \ Cut$$

The two upper cuts, on $w : A \supset B$, are of smaller derivation height, the other two on the smaller cut formulas $o : A, o : B$.   QED.

We obtain at once the result that each of the seven interpolable intermediate logics (*cf.* Maksimova 1979, Chagrov and Zakharyaschev 1997) belong to the class **G3I***. The point is simply that all these have frame conditions expressible geometrically:

1. **Int: Intuitionistic Logic.** The accessibility relation $\leqslant$ is reflexive and transitive, i.e.,

$$\forall w(w \leqslant w) \text{ and } \forall wor(w \leqslant o \& o \leqslant r \supset w \leqslant r).$$

2. **Jan: Jankov–De Morgan Logic** (cf. Jankov 1968). This logic, also known as **KC** (cf. Chagrov and Zakharyaschev 1997) and as the 'logic of weak excluded middle', is axiomatized by one of the formulas $\neg A \lor \neg\neg A$ or $\neg(A \& B) \supset \neg A \lor \neg B$.

   The relation $\leqslant$ is **directed** or **convergent**, i.e.,

$$\forall wor(w \leqslant o \& w \leqslant r \supset \exists l(o \leqslant l \& r \leqslant l)).$$

   The instance of the rule scheme generated by this frame condition is, with l fresh,

$$\frac{o \leqslant l, r \leqslant l, w \leqslant o, w \leqslant r, \Gamma \to \Delta}{w \leqslant o, w \leqslant r, \Gamma \to \Delta}$$

3. **GD: Gödel–Dummett Logic.** This logic (also known as **LC**, for 'linear chains') has as characteristic axiom scheme either $(A \supset B) \lor (B \supset A)$ or $((A \supset B) \supset C) \supset (((B \supset A) \supset C) \supset C)$.

   The accessibility relation is **strongly connected**, i.e.,

$$\forall wor(w \leqslant o \& w \leqslant r \supset o \leqslant r \lor r \leqslant o).$$

   The instance of the rule scheme generated by this frame condition is

$$\frac{o \leqslant r, w \leqslant o, w \leqslant r, \Gamma \to \Delta \quad r \leqslant o, w \leqslant o, w \leqslant r, \Gamma \to \Delta}{w \leqslant o, w \leqslant r, \Gamma \to \Delta}$$

4. **Bd$_2$:** This logic is axiomatized by, for example, $A \lor (A \supset (B \lor \neg B))$.

   The accessibility relation has **depth at most** 2, i.e., it satisfies

$$\forall wor(w \leqslant o \& o \leqslant r \supset r \leqslant o \lor o \leqslant w).$$

   The instance of the rule scheme generated by this frame condition is

$$\frac{r \leqslant o, w \leqslant o, o \leqslant r, \Gamma \to \Delta \quad o \leqslant w, w \leqslant o, o \leqslant r, \Gamma \to \Delta}{w \leqslant o, o \leqslant r, \Gamma \to \Delta}$$

5. **GSc:** The logic is axiomatized by $(A \supset B) \lor (B \supset A) \lor ((A \supset \neg B) \& (\neg B \supset A))$ and $A \lor (A \supset B \lor \neg B)$. The accessibility relation has depth at most 2 and at most 2 final elements, i.e., the following holds in

addition to the frame condition for **Bd$_2$**:

$$\forall wor \exists l((w \leqslant l \,\&\, o \leqslant l) \vee (o \leqslant l \,\&\, r \leqslant l) \vee (w \leqslant l \,\&\, r \leqslant l)).$$

The corresponding instantiation of the rule scheme for the first condition is given above; that for the second condition is, with l fresh:

$$\frac{w \leqslant l,\, o \leqslant l,\, \Gamma \to \Delta \quad o \leqslant l,\, r \leqslant l,\, \Gamma \to \Delta \quad w \leqslant l,\, r \leqslant l,\, \Gamma \to \Delta}{\Gamma \to \Delta}$$

6. **Sm: Smetanich logic**, also known as **LC$_2$** (cf. Chagrov and Zakharyaschev 1997) or the 'logic of here and there'. The accessibility relation is linear and has depth at most 2, i.e., satisfies the conditions for **GD** and **Bd$_2$**. It is axiomatized by the **GD** axiom plus the **Bd$_2$** axiom, or, equivalently, $(\neg B \supset A) \supset (((A \supset B) \supset A) \supset A)$.

7. **Cl: Classical logic.** The logic is axiomatized by $A \vee \neg A$ or by $\neg\neg A \supset A$. The accessibility relation is **symmetric**

$$\forall wo(w \leqslant o \supset o \leqslant w).$$

and the corresponding instantiation of the rule scheme is clearly

$$\frac{o \leqslant w,\, w \leqslant o,\, \Gamma \to \Delta}{w \leqslant o,\, \Gamma \to \Delta}$$

There are the following containments between these logics:

$$\textbf{Int} \subset \textbf{Jan} \subset \textbf{GD} \subset \textbf{Sm},\ \textbf{Int} \subset \textbf{Bd}_2 \subset \textbf{GSc} \subset \textbf{Sm} \text{ and } \textbf{Sm} \subset \textbf{Cl}.$$

We recall the standard translation $^\square$ of **Int** into **S4**, a variant from Troelstra and Schwichtenberg (2000) of the translation given in Gödel (1933):

$$P^\square \equiv \square P$$
$$\bot^\square \equiv \bot$$
$$(A \supset B)^\square \equiv \square(A^\square \supset B^\square)$$
$$(A \,\&\, B)^\square \equiv A^\square \,\&\, B^\square$$
$$(A \vee B)^\square \equiv A^\square \vee B^\square$$

The translation $\Gamma^\square$ of a multiset $\Gamma \equiv A_1, \ldots, A_n$ is defined componentwise by

$$(A_1, \ldots, A_n)^\square \equiv A_1^\square, \ldots, A_n^\square$$

The translation on relational atoms is the identity.

We obtain a uniform proof of the faithful embeddings of intermediate logics between **Int** and **Cl** and intermediate modal logics between **S4** and **S5**:

Theorem 12.31. *Given an extension* **G3I*** *of* **G3I** *with rules for* $\leqslant$, *let* **G3S4*** *be the corresponding extension of* **G3S4**. *Then* **G3I*** $\vdash \Gamma \to \Delta$ *if and only if* **G3S4*** $\vdash \Gamma^\square \to \Delta^\square$.

Proof. From left to right is routine, by induction on the structure of the derivation. For example, an initial sequent $w \leqslant o, \Gamma, w : P \to o : P, \Delta$ translates into the **G3S4*** derivation

$$\cfrac{\cfrac{\cfrac{\dots, \Gamma^\square, w : \square P, r : P \to r : P, \Delta^\square}{w \leqslant o, o \leqslant r, w \leqslant r, \Gamma^\square, w : \square P \to r : P, \Delta^\square} \; {\scriptstyle L\square}}{w \leqslant o, o \leqslant r, \Gamma^\square, w : \square P \to r : P, \Delta^\square} \; {\scriptstyle Trans}}{w \leqslant o, \Gamma^\square, w : \square P \to o : \square P, \Delta^\square} \; {\scriptstyle R\square}$$

Similarly, an instance of $R\supset$ is

$$\cfrac{w \leqslant o, \Gamma, o : A \to o : B, \Delta}{\Gamma \to w : A \supset B, \Delta} \; {\scriptstyle R\supset}$$

It translates into the steps

$$\cfrac{\cfrac{w \leqslant o, \Gamma^\square, o : A^\square \to o : B^\square, \Delta^\square}{w \leqslant o, \Gamma^\square \to o : A^\square \supset B^\square, \Delta^\square} \; {\scriptstyle R\supset}}{\Gamma^\square \to w : \square(A^\square \supset B^\square), \Delta^\square} \; {\scriptstyle R\square}$$

An instance of $L\supset$ is dealt with likewise. Conjunction, disjunction, and falsity are routine.

The converse direction follows from the following lemma:

Lemma 12.32. *If* Γ, Δ *are multisets of labelled formulas (with relational atoms also possibly in* Γ) *and* Γ', Δ' *are multisets of labelled atomic formulas, and* **G3S4*** $\vdash \Gamma^\square, \Gamma' \to \Delta^\square, \Delta'$, *then* **G3I*** $\vdash \Gamma, \Gamma' \to \Delta, \Delta'$.

Proof. By induction on the derivation of $\Gamma^\square, \Gamma' \to \Delta^\square, \Delta'$. If it is an initial sequent, then some atom $w : P$ is in Γ' and in Δ'; the conclusion then follows in **G3I*** by *Ref* from the initial sequent $w \leqslant w, \Gamma, \Gamma' \to \Delta, \Delta'$. If it is a conclusion of $L\perp$, so also is $\Gamma, \Gamma' \to \Delta, \Delta'$. If it is derived by a rule for & or for $\vee$, the inductive hypothesis applies to the premises and then the corresponding rule in **G3I*** gives the conclusion.

If it is derived by a modal rule, the principal formula, being a translated formula, can be only of the form $\square P$ or of the form $\square(A^\square \supset B^\square)$. There are thus four cases:

1. With $\square P$ principal on the left, the step

$$\cfrac{w \leqslant o, o : P, w : \square P, \Gamma''^\square, \Gamma' \to \Delta^\square, \Delta'}{w \leqslant o, w : \square P, \Gamma''^\square, \Gamma' \to \Delta^\square, \Delta'} \; {\scriptstyle L\square}$$

is translated to the admissible step in **G3I***

$$\frac{w \leqslant o, o : P, w : P, \Gamma'', \Gamma' \to \Delta, \Delta'}{w \leqslant o, w : P, \Gamma'', \Gamma' \to \Delta, \Delta'}$$

2. With $\Box P$ principal on the right, the step (with o fresh)

$$\frac{w \leqslant o, \Gamma^\Box, \Gamma' \to \Delta''^\Box, \Delta', o : P}{\Gamma^\Box, \Gamma' \to \Delta''^\Box, \Delta', w : \Box P} \ R\Box$$

is translated (using admissibility of substitution) to the steps in **G3I***

$$\frac{\dfrac{w \leqslant o, \Gamma, \Gamma' \to \Delta'', \Delta', o : P}{w \leqslant w, \Gamma, \Gamma' \to \Delta'', \Delta', w : P} \ (w/o)}{\Gamma, \Gamma' \to \Delta'', \Delta', w : P} \ \textit{Refl}$$

3. With $\Box(A^\Box \supset B^\Box)$ principal on the left, the step

$$\frac{w \leqslant o, w : \Box(A^\Box \supset B^\Box), o : A^\Box \supset B^\Box, \Gamma''^\Box, \Gamma' \to \Delta^\Box, \Delta'}{w \leqslant o, w : \Box(A^\Box \supset B^\Box), \Gamma''^\Box, \Gamma' \to \Delta^\Box, \Delta'} \ L\Box$$

gives, by height-preserving invertibility of $L\supset$ in **G3S4***, derivations in **G3S4*** of the sequents

$$w \leqslant o, w : \Box(A^\Box \supset B^\Box), \Gamma''^\Box, \Gamma' \to \Delta^\Box, \Delta', o : A^\Box$$

and

$$w \leqslant o, w : \Box(A^\Box \supset B^\Box), o : B^\Box, \Gamma''^\Box, \Gamma' \to \Delta^\Box, \Delta'$$

to which the inductive hypothesis applies. This gives us derivations in **G3I*** of the sequents

$$w \leqslant o, w : A \supset B, \Gamma'', \Gamma' \to \Delta, \Delta', o : A$$

and

$$w \leqslant o, w : A \supset B, o : B, \Gamma'', \Gamma' \to \Delta, \Delta'$$

from which the desired conclusion

$$w \leqslant o, w : A \supset B, \Gamma'', \Gamma' \to \Delta, \Delta'$$

follows by a step of $L\supset$ in **G3I***.

4. If $\Box(A^\Box \supset B^\Box)$ is principal on the right, the step is

$$\frac{w \leqslant o, \Gamma^\Box, \Gamma' \to \Delta''^\Box, \Delta', o : A^\Box \supset B^\Box}{\Gamma^\Box, \Gamma' \to \Delta''^\Box, \Delta', w : \Box(A^\Box \supset B^\Box)} \ R\Box$$

from which, by height-preserving invertibility of $R\supset$ in **G3S4***, we have a derivation in **G3S4*** of

$$w \leqslant o, o : A^{\square}, \Gamma^{\square}, \Gamma' \to \Delta''^{\square}, \Delta', o : B^{\square}$$

to which the inductive hypothesis applies. A step of $R\supset$ in **G3I*** gives us the desired conclusion.   QED.

Observe that the translation does not affect the steps involving the rules for the accessibility relation; therefore the faithfulness of the embedding is maintained for each of the intermediate logics considered above and even for those not considered, provided the frame conditions are geometric implications.

Observe also that the admissibility of *Contraction* and *Cut* in **G3I*** may be obtained from this result, and their admissibility for extensions of **S4**, since no use is made thereof in the proof of the theorem.

One may conclude, therefore, in an easy uniform fashion, the faithfulness of the embedding of each intermediate logic characterized by frames satisfying geometric implications into its (smallest) modal companion. Well-known modal companions are **S4** for **Int**, **S4.2** for **Jan**, **S4.3** for **GD**, **S5** for **Cl**.

12.4 Substructural logics

From the point of view of proof theory as understood in this book, what are called substructural logics can be described as follows. Consider a system of sequent calculus in which some structural rule such as weakening or contraction is indispensable. The **G0i**-calculus of section 5.1 in *Structural Proof Theory* is an example. What happens to derivability in the calculus if one rule, say weakening, is left out? The resulting logical system can be characterized also axiomatically and is known as **relevant logic**. One oft-heard characterization of substructural logics is that they are 'logics without structural rules'. Note, however, that the absence or presence of structural rules is a property of a logical calculus, not of a logic in itself. Were this not so, all the calculi of the **G3**-class would count as calculi for substructural logics, even if they include intuitionistic and classical logic. Thus, the terminology is somewhat misleading.

The family of relevant, and, more generally, substructural logics, is among the logics that can be characterized in terms of a relational semantics. Our method can therefore be successfully applied to obtain sequent calculi for these logics. The starting point for the development of uniform calculi for substructural logics is the **Routley-Meyer** relational semantics. This semantics is a generalization of the standard relational semantics for intuitionistic and modal logic: instead of a binary accessibility relation, we have a ternary relation R on a set of worlds W. A distinguished element 0 of W defines a projection of R, namely $a \leqslant b \equiv R0ab$ that turns out to be a partial order.

For basic relevant logic, R satisfies the following properties of reflexivity and componentwise monotonicity:

Ref $R0ww$

Mon_1 $R0w'w \,\&\, Rwor \supset Rw'or$

Mon_2 $R0o'o \,\&\, Rwor \supset Rwo'r$

Mon_3 $R0r'r \,\&\, Rwor' \supset Rwor$

All the above properties can be given as rules for the accessibility relation to be added to a suitable labelled calculus.

As for intuitionistic logic, the only connective with a non-trivial semantics is implication, with validity defined inductively:

$w \Vdash A \supset B$ *whenever for all o, r, from $Rwor$ and $o \Vdash A$ follows $r \Vdash A$.*

This semantic explanation justifies the rules

$$\frac{Rwor, w : A \supset B, \Gamma \to \Delta, o : A \quad Rwor, w : A \supset B, r : B, \Gamma \to \Delta}{Rwor, w : A \supset B, \Gamma \to \Delta} \; L\supset$$

$$\frac{Rwor, o : A, \Gamma \to \Delta, r : B}{\Gamma \to \Delta, w : A \supset B} \; R\supset$$

The latter has the variable condition $o, r \notin \Gamma, \Delta, w:A \supset B$.

These rules give a cut-free complete sequent calculus for basic relevance logic, with initial sequents given by

$$R0wo, w : P, \Gamma \to \Delta, o : P.$$

The logical rules for implication are given as above, the rules for & and ∨ as in **G3K** and **G3I**, and the mathematical rules are given by the monotonicity properties of R.

Besides cut, also the other structural rules, namely weakening and contraction, are admissible. We observe that this does not contradict the substructural nature of these logics. These admissible rules are what could be called, borrowing terminology from hypersequents, 'external' structural rules. In fact, we can easily verify that the axiom $A \supset (B \supset A)$ that corresponds to weakening is not derivable in the above system despite the admissibility of weakening.

Logics that extend the basic relevant logic can be obtained by assuming additional properties for the accessibility relation. We recall some correspondences between axioms and frame properties for a variety of relevant logics. First, define $R^2abcd \equiv R^2(ab)cd \equiv \exists w(Rabw \& Rwcd)$ and $R^2a(bc)d \equiv \exists w(Rawd \& Rbcw)$:

Table 12.4 Axioms and frame properties for substructural logics

Axiom	Frame property
$A\&(A \supset B) \supset B$	$Raaa$ or $R0ab \supset Raab$ idempotence
$(A \supset B)\&(B \supset C) \supset (A \supset C)$	$Rabc \supset R^2a(ab)c$ transitivity
$(A \supset B) \supset ((B \supset C) \supset (A \supset C))$	$R^2abcd \supset R^2b(ac)d$ suffixing
$(A \supset B) \supset ((C \supset A) \supset (C \supset B))$	$R^2abcd \supset R^2a(bc)d$ associativity
$(A \supset (A \supset B)) \supset (A \supset B)$	$Rabc \supset R^2abbc$ contraction
$((A \supset A) \supset B) \supset B$	$Ra0a$ specialized assertion
$A \supset ((A \supset B) \supset B)$	$Rabc \supset Rbac$ commutativity
$A \supset (A \supset A)$	$Rabc \supset (R0ac \lor R0bc)$ mingle

All the properties of R are given by geometric implications. As a consequence, the basic calculus can be extended by rules that represent these frame properties. The structural properties follow from the general result on extensions with the geometric rule scheme.

Notes to Chapter 12

Section 12.2: A rule similar to $\Box R\text{-}L$, but in natural deduction style, is given in Gabbay (1996).

Semantic proofs of closure with respect to cut for certain sequent calculi for **GL**, based on completeness arguments, were presented in Sambin and Valentini (1982) and in Avron (1984). Syntactic proofs, aimed at providing explicit proof transformations that would describe a procedure of cut elimination, were proposed by Leivant (1981), Valentini (1983), and Borga (1983). Valentini (1983) gave a counterexample to the proof presented by Leivant. More recently Moen (2003) observed that the proof by Valentini assumes as a starting point a reduction of a cut on $\Box A$ to a detour cut, which is not fully justified in a calculus with explicit contraction. However, in all the proofs given in the 1980s (and also in more recent proposals; see Sasaki 2002) calculi with contexts-as-sets have been used, but these are not altogether satisfactory, as discussed in Section 6.1(b).

Another problematic aspect of the proposed calculi for provability logic is a lack of harmony; in fact, there is only one rule, acting on both left and right, for $\Box$

$$\frac{\Box\Gamma, \Gamma, \Box A \to A}{\Box\Gamma, \Gamma' \to \Delta, \Box A}$$

The rule does not respect any of the principles of separation, symmetry, and uniqueness put forward as good design requirements for sequent calculi in Wansing (1994). A discussion of the notion of harmony in the context of modal logic is presented in Read (2008). The general elimination rules of natural deduction aim at such harmony, and in von Plato (2005) they are used to give a solution to the problem of normal derivability in **S4**.

Here we have shown how a calculus with admissible contraction for sequents labelled by possible worlds, with harmonious, semantically originated left and right rules for $\Box$, permits a transparent syntactic proof of cut elimination for **GL**.

A recent proof of cut elimination for **GL** appears in Goré and Ramanayake (2008). Using an argument from von Plato (2001b), they show how to tackle the problematic case that arises in Valentini's proof if an explicit rule of contraction is used in place of the implicit 'context-as-sets' treatment.

Subsection 12.3(a) is based on Dyckhoff and Negri (2005). A translation $(\cdot)^*$ from the language of intuitionistic propositional logic to the language of classical modal logic was defined by Gödel in 1933. He proved by induction on derivations that his translation was *sound*, that is, if $\vdash_{Int} A$, then $\vdash_{S4} A^*$, and conjectured faithfulness of the embedding, i.e. the converse. This was proved by McKinsey and Tarski (1948), who gave a semantic proof that $\nvdash_{Int} A$ implies $\nvdash_{S4} A^*$. Dummett and Lemmon (1959) proved, using the same semantic method, that $\vdash_{Int+Ax} A$ if and only if $\vdash_{S4+Ax^*} A^*$ where A is any propositional formula and Ax is a collection of axioms.

Compared with a standard proof for unlabelled calculi (Troelstra and Schwichtenberg 2000), the above is both simple and general. The core of the above proof, that is, the erasure of each $\Box$, is reminiscent of an analogous reduction in the model-theoretic proof of faithfulness of the embedding of **Int** into **S4**. For that purpose, it is shown how a countermodel for an unprovable sequent in **Int** is turned into a countermodel for the translation of that sequent in **S4**; in particular, 'it can be treated as a modal frame isomorphic to its skeleton' (see Theorem 3.83 in Chagrov and Zakharyaschev 1997).

Subsection 12.3(b): A similar approach to substructural logics is presented in Viganó (2000). The main difference with respect to our method consists in the use of a basic sequent calculus with explicit structural rules and in a presentation of mathematical rules for the accessibility relation in the form of rules with a single conclusion (Horn clauses) that cannot be extended beyond Harrop theories (theories that do not have disjunctions in positive parts of axioms). This excludes, for instance, the treatment of the last frame property in Table 12.4.

For a general background, history, motivations, applications, and references to the vast literature in the field of substructural logics, we refer to the survey by Dunn and Restall (2002) and to the two recent monographs Restall (2000) and Mares (2004).

Bibliography

Avron, A. (1984) On modal systems having arithmetical interpretations. *The Journal of Symbolic Logic,* vol. 49, pp. 935–942.

Basin, D., S. Matthews, and L. Viganó (1998) Natural deduction for non-classical logics. *Studia Logica,* vol. 60, pp. 119–160.

van Benthem, J. (1984) Correspondence theory. In D. Gabbay and F. Guenther (eds.) *Handbook of Philosophical Logic,* vol. 2, pp. 167–247, Kluwer, Dordrecht.

Bernays, P. (1945) Review of Ketonen (1944). *The Journal of Symbolic Logic,* vol. 10, pp. 127–130.

Bezem, M. and D. Hendriks (2008) On the mechanization of the proof of Hessenberg's theorem in coherent logic. *Journal of Automated Reasoning,* vol. 40, pp. 61–85.

Blackburn, P. (2000) Representation, reasoning, and relational structures: a hybrid logic manifesto. *Logic Journal of the IGPL,* vol. 8, pp. 339–365.

Blackburn, P., M. de Rijke, and Y. Venema (2001) *Modal Logic.* Cambridge University Press.

Blass, A. (1988) Topoi and computation. *Bulletin of the European Association for Theoretical Computer Science,* vol. 36, pp. 57–65.

Borga, M. (1983) On some proof theoretical properties of the modal logic GL. *Studia Logica,* vol. 4, pp. 453–459.

Braüner, T. (2000) A cut-free Gentzen formulation of the modal logic S5. *Logic Journal of the IGPL,* vol. 8, pp. 629–643.

Brouwer, L. (1924) Intuitionistische Zerlegung mathematischer Grundbegriffe. As reprinted in Brouwer's *Collected Works,* vol. 1, pp. 275–280, North-Holland, Amsterdam, 1980.

(1927) Virtuelle Ordnung und unerweiterbare Ordnung. As reprinted in Brouwer's *Collected Works,* vol. 1, pp. 406–408.

(1950) Remarques sur la notions d'ordre. As reprinted in Brouwer's *Collected Works,* vol. 1, pp. 499–500.

Bull, R. and K. Segerberg (1984) Basic modal logic. In D. Gabbay and F. Guenther (eds.) *Handbook of Philosophical Logic,* vol. 2, pp. 1–88, Kluwer, Dordrecht. Second edition 2001.

Burris, S. (1995) Polynomial time uniform word problems. *Mathematical Logic Quarterly,* vol. 41, pp. 173–182.

Buss, S. (1995) On Herbrand's theorem. In D. Leivant (ed.) *Logical and Computational Complexity* (LNCS 960), pp. 195–209, Springer, Berlin.

Castellini, C. (2005) *Automated Reasoning in Quantified Modal and Temporal Logic.* PhD thesis, School of Informatics, University of Edinburgh.

Castellini, C. and A. Smaill (2002) A systematic presentation of quantified modal logics. *Logic Journal of the IGPL*, vol. 10, pp. 571–599.

Catach, L. (1991) Tableaux: a general theorem prover for modal logics. *Journal of Automated Reasoning*, vol. 7, pp. 489–510.

Cederqvist, J., T. Coquand, and S. Negri (1998) The Hahn-Banach theorem in type theory. In G. Sambin and J. Smith (eds.) *Twenty-Five Years of Constructive Type Theory*, pp. 39–50, Oxford University Press.

Chagrov, A. and M. Zakharyaschev (1997) *Modal Logic.* Oxford University Press.

Copeland, B. J. (2002) The genesis of possible worlds semantics. *Journal of Philosophical Logic*, vol. 31, pp. 99–137.

Coste, M., H. Lombardi, and M.-F. Roy (2001) Dynamical method in algebra: effective Nullstellensätze. *Annals of Pure and Applied Logic*, vol. 111, pp. 203–256.

Curry, H. B. (1952) The elimination theorem when modality is present. *The Journal of Symbolic Logic*, vol. 17, pp. 249–265.

van Dalen, D. and R. Statman (1978) Equality in the presence of apartness. In J. Hintikka *et al.* (eds.) *Essays in Mathematical and Philosophical Logic*, pp. 95–116, D. Reidel, Dordrecht.

Degtyarev, A. and A. Voronkov (2001) Equality reasoning in sequent-based calculi. In A. Robinson and A. Voronkov (eds.) *Handbook of Automated Reasoning*, vol. 1, pp. 611–706, Elsevier, Amsterdam.

Dragalin, A. (1988) *Mathematical Intuitionism: Introduction to Proof Theory.* American Mathematical Society, Providence, RI.

Dummett, M. (1959) A propositional calculus with a denumerable matrix. *The Journal of Symbolic Logic*, vol. 24, pp. 97–106.

Dummett, M. and E. Lemmon (1959) Modal logics between S4 and S5. *Zeitschrift für Mathematische Logik und Grundlagen der Mathematik*, vol. 5, pp. 250–264.

Dunn, J. M. and G. Restall (2002) Relevance logic. In D. Gabbay and F. Guenthner (eds.) *Handbook of Philosophical Logic*, vol. 6, pp. 1–128, Kluwer, Dordrecht.

Dyckhoff, R. (1988) Implementing a simple proof assistant. In J. Derrick and H. A. Lewis (eds.) *Workshop on Programming for Logic Teaching*, pp. 49–59. Proceedings 23/1988, University of Leeds: Centre for Theoretical Computer Science. Mimeograph.

Dyckhoff, R. and S. Negri (2005) Proof analysis in intermediate propositional logics. Ms.

(2006) Decision methods for linearly ordered Heyting algebras. *Archive for Mathematical Logic*, vol. 45, pp. 411–422.

Ehrenfeucht, A. (1959) Decidability of the theory of linear ordering relation. *Notices of the American Mathematical Society*, vol. 6, pp. 268–269.

Fitch, F. (1966) Tree proofs in modal logic. *The Journal of Symbolic Logic*, vol. 31, p. 152.

Fitting, M. (1983) *Proof Methods for Modal and Intuitionistic Logics*. D. Reidel, Dordrecht.

(1999) A simple propositional S5 tableau system. *Annals of Pure and Applied Logic*, vol. 96, pp. 107–115.

Fitting, M. and R. L. Mendelsohn (1998) *First-Order Modal Logic*. Synthese Library vol. 277. Kluwer, Dordrecht.

Freese, R., J. Ježek, and J. Nation (1995) *Free Lattices*. American Mathematical Society, Providence, RI.

Gabbay, D. (1996) *Labelled Deductive Systems*. Oxford University Press.

Gentzen, G. (1934–35) Untersuchungen über das logische Schließen. *Mathematische Zeitschrift*, vol. 39, pp. 176–210 and 405–431.

(1938) Neue Fassung des Widerspruchsfreiheitsbeweises für die reine Zahlentheorie. *Forschungen zur Logik und zur Grundlegung der exakten Wissenschaften*, vol. 4, pp. 19–44.

(1969) *The Collected Papers of Gerhard Gentzen*. Ed. M. Szabo, North-Holland, Amsterdam.

(2008) The normalization of derivations. *The Bulletin of Symbolic Logic*, vol. 14, pp. 245–257.

Girard, J.-Y. (1987) *Proof Theory and Logical Complexity*, vol. 1. Bibliopolis, Naples.

Gödel, K. (1933) Eine Interpretation des intuitionistischen Aussagenkalküls. *Ergebnisse eines mathematischen Kolloquiums*, vol. 4, pp. 39–40. English tr. in Gödel's *Collected Works* (1986), vol. 1, pp. 300–303.

Goldblatt, R. (2005) Mathematical modal logic: a view of its evolution. In D. Gabbay and J. Woods (eds.) *Handbook of the History of Logic*, vol. 7, pp. 1–98, Elsevier, Amsterdam.

Goré, R. (1998) Tableau methods for modal and temporal logis. In M. D'Agostino *et al.* (eds.) *Handbook of Tableau Methods*, Kluwer, Dordrecht.

Goré, R. and R. Ramanayake (2008) Valentini's cut-elimination for provability logic resolved. In C. Areces and R. Goldblatt (eds.) *Advances in Modal Logic*, vol. 7, pp. 67–96, College Publications, London.

Hallett, M. and U. Majer (ed.) (2004) *David Hilbert's Lectures on the Foundations of Geometry 1891–1902*. Springer, Berlin.

Hakli, R. and S. Negri (2011a) Does the deduction theorem fail for modal logic? *Synthese*, in press.

(2011b) Reasoning about collectively accepted group beliefs. *Journal of Philosophical Logic*, vol. 40, pp. 531–555.

Heyting, A. (1927) Zur intuitionistischen Axiomatik der projektiven Geometrie. *Mathematische Annalen*, vol. 98, pp. 491–538.

(1956) *Intuitionism, an Introduction*, North-Holland, Amsterdam.

Hilbert, D. (1899) *Grundlagen der Geometrie*. Teubner, Leipzig.

Janisczak, A. (1953) Undecidability of some simple formalized theories. *Fundamenta Mathematicae*, vol. 40, pp. 131–139.

Jankov, V. A. (1968) The calculus of the weak 'law of excluded middle'. *Mathematics of the USSR: Izvestija*, vol. 2, pp. 997–1004.

Joachimski, F. and R. Matthes (2003) Short proofs of normalization for the simply typed λ-calculus, permutative conversions and Gödel's T. *Archive for Mathematical Logic*, vol. 42, pp. 59–87.

Johnstone, Peter T. (1977) *Topos Theory*. LMS Monograph no. 10. Academic Press, London.

Kanger, S. (1957) *Provability in Logic*. Almqvist & Wiksell, Stockholm.

　(1963) A simplified proof method for elementary logic. In P. Braffort and D. Hirshberg (eds.) *Computer Programming and Formal Systems*, pp. 87–94, North-Holland, Amsterdam.

Ketonen, O. (1943) 'Luonnollisen päättelyn' kalkyylista (On the calculus of 'natural deduction', in Finnish). *Ajatus* (Yearbook of the Finnish Philosophical Society), vol. 12, pp. 128–140.

　(1944) *Untersuchungen zum Prädikatenkalkül*. Ann. Acad. Sci. Fenn., Ser. A.I. 23.

Kleene, S. (1952) *Introduction to Metamathematics*. North-Holland, Amsterdam.

Kohlenbach, U. (2008) *Applied Proof Theory: Proof Interpretations and Their Use in Mathematics*. Springer, Berlin.

Kreisel, G. (1954) Review of Janisczak (1953). *Mathematical Reviews*, vol. 15, pp. 669–670.

Kripke, S. (1959) A completeness theorem in modal logic. *The Journal of Symbolic Logic*, vol. 24, pp. 1–14.

　(1963a) Semantical analysis of modal logic I. Normal modal propositional calculi. *Zeitschrift für mathematische Logik und Grundlagen der Mathematik*, vol. 9, pp. 67–96.

　(1963b) Semantical considerations on modal logic. *Acta Philosophica Fennica*, vol. 16, pp. 83–94.

　(1965) Semantical Analysis of Intuitionistic Logic I. In M. Dummett and J. Crossley (eds.) *Formal Systems and Recursive Functions*, pp. 92–130, North-Holland, Amsterdam.

Kushida, H. and M. Okada (2003) A proof-theoretic study of the correspondence of classical logic and modal logic. *The Journal of Symbolic Logic*, vol. 68, pp. 1403–1414.

Läuchli, H. and J. Leonard (1966) On the elementary theory of linear order. *Fundamenta Mathematicae*, vol. 59, pp. 109–116.

Leivant, D. (1981) On the proof theory of the modal logic for arithmetic provability. *The Journal of Symbolic Logic*, vol. 46, pp. 531–538.

Lopez-Escobar, L. (1999) Standardizing the N-systems of Gentzen. In X. Caicedo and C. Montenegro (eds.) *Models, Algebras and Proofs*, pp. 411–434, Dekker, New York.

McKinsey, J. and A. Tarski (1948) Some theorems about the sentential calculi of Lewis and Heyting. *The Journal of Symbolic Logic*, vol. 13, pp. 1–15.

Mac Lane, S. and I. Moerdijk (1992) *Sheaves in Geometry and Logic*. Springer, New York.

Maksimova, L. (1979) Interpolation properties of superintuitionistic logics. *Studia Logica*, vol. 38, pp. 419–428.

Mares, E. (2004) *Relevant logic: A Philosophical Interpretation*. Cambridge University Press.

Massacci, F. (2000) Single step tableaux for modal logics: methodology, computations, algorithms. *Journal of Automated Reasoning*, vol. 24, pp. 319–364.

Meinander, A. (2010) A solution of the uniform word problem for ortholattices. *Mathematical Structures in Computer Science*, vol. 20, pp. 625–638.

Mints, G. (1970) Cut-free calculi of type S5. Translation of Russian original in A. Slisenko (ed.) *Studies in Constructive Mathematics and Mathematical Logic*, vol. 8, pt II, pp. 79–82, n.p., Leningrad.

Mints, G. (1997) Indexed systems of sequents and cut-elimination. *Journal of Philosophical Logic*, vol. 26, pp. 671–696.

Moen, A. (2003) A normal form for Gödel Löb sequent calculus, manuscript. Abstract published in *The Bulletin of Symbolic Logic*, vol. 10 (2004), p. 266.

Mostowski, A. (1965) *Thirty Years of Foundational Studies*. The Philosophical Society of Finland, Helsinki.

Negri, S. (1999) Sequent calculus proof theory of intuitionistic apartness and order relations. *Archive for Mathematical Logic*, vol. 38, pp. 521–547.

(2003) Contraction-free sequent calculi for geometric theories, with an application to Barr's theorem. *Archive for Mathematical Logic*, vol. 42, pp. 389–401.

(2005a) Proof analysis in modal logic. *Journal of Philosophical Logic*, vol. 34, pp. 507–544.

(2005b) Permutability of rules for linear lattices. *Journal of Universal Computer Science*, vol. 11, pp. 1986–1995.

(2008) Proof analysis in non-classical logics. In C. Dimitracopoulos, L. Newelski, D. Normann, and J. Steel (eds.) *Logic Colloquium '05*, ASL Lecture Notes in Logic, vol. 28, pp. 107–128, Cambridge University Press.

(2009) Kripke completeness revisited. In G. Primiero and S. Rahman (eds.) *Acts of Knowledge – History, Philosophy and Logic*, pp. 247–282, College Publications, London.

Negri, S. and J. von Plato (1998) Cut elimination in the presence of axioms. *The Bulletin of Symbolic Logic*, vol. 4, pp. 418–435.

(2001) *Structural Proof Theory*. Cambridge University Press.

(2002) Permutability of rules in lattice theory. *Algebra Universalis*, vol. 48, pp. 473–477.

(2004) Proof systems for lattice theory. *Mathematical Structures in Computer Science*, vol. 14, pp. 507–526.

(2005) The duality of classical and constructive notions and proofs. In L. Crosilla and P. Schuster (eds.) *From Sets and Types to Topology and Analysis: Towards*

Practicable Foundations for Constructive Mathematics, pp. 149–161, Oxford University Press.

Negri, S., J. von Plato, and Th. Coquand (2004) Proof-theoretic analysis of order relations. *Archive for Mathematical Logic*, vol. 43, pp. 297–309.

Nerode, A. (1991) Some lectures on modal logic. In F. Bauer (ed.) *Logic, Algebra, and Computation*, NATO ASI Series, Springer, New York.

Ohlbach, H. (1993) Translation methods for non-classical logics – an overview. *Bulletin of the IGPL*, vol. 1, pp. 69–91.

Ohnishi, M. and K. Matsumoto (1957) Gentzen method in modal calculi. *Osaka Mathematical Journal*, vol. 9, pp. 113–130.

Palmgren, E. (2002) An intuitionistic axiomatization of real closed fields. *Mathematical Logic Quarterly*, vol. 48, pp. 297–299.

von Plato, J. (1995) The axioms of constructive geometry. *Annals of Pure and Applied Logic*, vol. 76, pp. 169–200.

(1996) Organization and development of a constructive axiomatization. In S. Berardi and M. Coppo (eds.) *Types for Proofs and Programs* (LNCS 1158), pp. 288–296, Springer, Berlin.

(1997) Formalization of Hilbert's geometry of incidence and parallelism. *Synthese*, vol. 110, pp. 127–141.

(2001a) Natural deduction with general elimination rules. *Archive for Mathematical Logic*, vol. 40, pp. 541–567.

(2001b) A proof of Gentzen's *Hauptsatz* without multicut. *Archive for Mathematical Logic*, vol. 40, pp. 9–18.

(2001c) Positive lattices. In P. Schuster *et al.* (eds.) *Reuniting the Antipodes*, pp. 185–197, Kluwer, Dordrecht.

(2004) Ein Leben, ein Werk – Gedanken über das wissenschaftliche Schaffen des finnischen Logikers Oiva Ketonen. In R. Seising (ed.) *Form, Zahl, Ordnung: Studien zur Wissenschafts- und Technikgeschichte*, pp. 427–435, Franz Steiner Verlag, Stuttgart.

(2005) Normal derivability in modal logic. *Mathematical Logic Quarterly*, vol. 51, pp. 632–638.

(2007) In the shadows of the Löwenheim-Skolem theorem: early combinatorial analyses of mathematical proofs. *The Bulletin of Symbolic Logic*, vol. 13, pp. 189–225.

(2008) Gentzen's proof of normalization for intuitionistic natural deduction. *The Bulletin of Symbolic Logic*, vol. 14, pp. 240–244.

(2009) Gentzen's logic. In D. Gabbay and J. Woods (eds.) *Handbook of the History of Logic*, vol. 5: *Logic from Russell to Church*, pp. 667–721, Elsevier, Amsterdam.

(2010) Combinatorial analysis of proofs in projective and affine geometry. *Annals of Pure and Applied Logic*, vol. 162, no. 2, pp. 144–161.

Pohlers, W. (2009) *Proof Theory: The First Step into Impredicativity*. Springer, Berlin.

Popkorn, S. [Harold Simmons] (1994) *First Steps in Modal Logic*. Cambridge University Press.

Prawitz, D. (1965) *Natural Deduction: A Proof-Theoretical Study*. Almqvist & Wicksell, Stockholm.

(1971) Ideas and results in proof theory. In J. Fenstad (ed.) *Proceedings of the Second Scandinavian Logic Symposium*, pp. 235–308, North-Holland, Amsterdam.

Read, S. (2008) Harmony and Modality. In C. Dégremont, L. Kieff, and H. Rückert (eds.) *Dialogues, Logics and Other Strong Things: Essays in Honour of Shahid Rahman*, pp. 285–303, College Publications, London.

Restall, G. (2000) *An Introduction to Substructural Logics*. Routledge.

(2008) Proofnets for S5: sequents and circuits for modal logic. In C. Dimitracopoulos, L. Newelski, D. Normann, and J. Steel (eds.) *Logic Colloquium '05*, ASL Lecture Notes in Logic, vol. 28, pp. 151–172, Cambridge University Press.

Sambin, G. and S. Valentini (1982) The modal logic of provability: the sequential approach. *Journal of Philosophical Logic*, vol. 11, pp. 311–342.

Sasaki, K. (2002) A cut-free sequent system for the smallest interpretability logic. *Studia Logica*, vol. 70, pp. 353–372.

Sato, M. (1980) A cut-free Gentzen-type system for the modal logic S5. *The Journal of Symbolic Logic*, vol. 45, pp. 67–84.

Schmidt, R. and U. Hustadt (2003) A principle for incorporating axioms into the first-order translation of modal formulae. *Lecture Notes in Artificial Intelligence*, vol. 2741, pp. 412–426.

Schroeder-Heister, P. (1984) A natural extension of natural deduction. *The Journal of Symbolic Logic*, vol. 49, pp. 1284–1300.

Scott, D. (1968) Extending the topological interpretation to intuitionistic analysis. *Compositio Mathematica*, vol. 20, pp. 194–210.

Shvarts, G. (1989) Gentzen style systems for K45 and K45D. In A. R. Meyer and M. A. Taitslin (eds.) *Logic at Botik '89* (LNCS 363), pp. 245–256, Springer, Berlin.

Simpson, A. (1994) *Proof Theory and Semantics of Intuitionistic Modal Logic*. PhD thesis, School of Informatics, University of Edinburgh.

Skolem, T. (1920) Logisch-kombinatorische Untersuchungen über die Erfüllbarkeit oder Beweisbarkeit mathematischer Sätze, nebst einem Theoreme über dichte Mengen. As reprinted in Skolem 1970, pp. 103–136.

(1970) *Selected Papers in Logic*. Ed. J. Fenstad, University of Oslo Press, Oslo.

Smorynski, C. (1973) Elementary intuitionistic theories. *The Journal of Symbolic Logic*, vol. 38, pp. 102–134.

Solovay, R. (1976) Provability interpretations of modal logic. *Israel Journal of Mathematics*, vol. 25, pp. 287–304.

Stouppa, P. (2007) A deep inference system for the modal logic S5. *Studia Logica*, vol. 85, pp. 199–214.

Szpilrajn, E. (1930) Sur l'extension de l'ordre partiel. *Fundamenta Mathematicae*, vol. 16, pp. 386–389.

Takeuti, G. (1987) *Proof Theory*. North-Holland, Amsterdam.

Tarski, A. (1949) On essential undecidability. *The Journal of Symbolic Logic*, vol. 14, pp. 75–76.

Tennant, N. (1992) *Autologic*. Edinburgh University Press.

Troelstra, A. and D. van Dalen (1988) *Constructivism in Mathematics*, vol. 1. North-Holland, Amsterdam.

Troelstra, A. and H. Schwichtenberg (1996) *Basic Proof Theory*. Cambridge University Press. Second edition 2000.

Valentini, S. (1983) The modal logic of provability: cut-elimination. *Journal of Philosophical Logic*, vol. 12, pp. 471–476.

Vickers, S. (1989) *Topology via Logic*. Cambridge University Press.

Viganó, L. (2000) *Labelled Non-Classical Logics*. Kluwer, Dordrecht.

Wang, H. (1960) Towards mechanical mathematics. *IBM Journal of Research and Development*, vol. 4, pp. 2–22.

Wansing, H. (1994) Sequent calculi for normal modal propositional logics. *Journal of Logic and Computation*, vol. 4, pp. 125–142.

(ed.) (1996) *Proof Theory of Modal Logic*. Kluwer, Dordrecht.

(2002) Sequent systems for modal logics. In D. Gabbay and F. Guenther (eds.) *Handbook of Philosophical Logic*, 2nd edn, vol. 8, pp. 61–145, Kluwer, Dordrecht.

Whitman, P. (1941) Free lattices. *Annals of Mathematics*, vol. 42, pp. 325–330.

Index of names

Avron A. 251

Barcan, R. 224, 225
Barr, M. 13, 144, 145
Basin, D. 221
Bayart, A. 187
van Benthem, J. 211
Bernays, P. 1, 3, 4, 5, 181
Beth, E. 213
Bezem, M. 181
Blackburn, P. 211, 221
Blass, A. 145
von Boguslawski, M. x
Boretti, B. x
Borga, M. 251
Braüner, T. 220
Brouwer, L. 18, 42, 148, 156
Bull, R. 220
Burris, S. 49, 82
Buss, S. 112

Cantor, G. 1
Carnap, R. 187
Castellini, C. 220
Catach, L. 220
Cederquist, J. 119
Chagrov, A. 244–246, 252
Copeland, B. 221
Coquand, T. 11, 119, 128
Coste, M. 145
Curry, H. 220

van Dalen, D. 38, 145
Dedekind, R. 45
Degtyarev, A. 112
Desargues, R. 181
Dragalin, A. 111, 145
Dummett, M. 129, 245, 252
Dunn, J. 252
Dyckhoff, R. x, 38, 129, 252

Ehrenfeucht, A. 129
Euclid 39, 42, 149

Fitch, F. 221
Fitting, M. 220, 224
Freese, R. 82
Frege, G. 3, 5, 6
Friedman, H. 145

Gabbay, D. 221, 251
Gentzen, G. 1–8, 10, 11, 17, 18, 23, 25, 26, 37, 38, 47, 85, 90, 92, 96, 104, 107, 111, 181, 220,
Girard, J.-Y. 97
Gödel, K. 187, 234, 245, 246, 252
Goldblatt, R. 187, 221
Gore, R. 220, 252
Grothendiek, A. 145

Hakli, R. 219
Hallett, M. 49
Harrop, R. 29, 55, 82, 114, 116, 119, 120, 150, 253
Hendriks, D. 181
Henkin, L. 213, 221
Herbrand, J. 13, 14, 82, 110, 112, 147, 153, 154
Heyting, A. 18, 42, 129, 148, 156
Hilbert, D. ix, x, 1, 3–6, 39–41, 49,
Hintikka, J. 187
Horn, A. 252
Hustadt, U. 221

Janisczak, A. 128
Jankov, V. 245
Jezek, J. 82
Joachimski, F. 29
Johnstone, P. 145

Kanckos, A. x
Kanger, S. 112, 187, 220
Ketonen, J. xi
Ketonen, O. 111, 181
Kleene, S. 99, 111
Kohlenbach, U. x
Kolmogorov, A. 18

Index of subjects